THE SILENT PULSE

THE SILENT PULSE

A Search for the Perfect Rhythm That Exists in Each of Us

George Leonard

E. P. DUTTON • NEW YORK

For information contact:

E. P. Dutton, 2 Park Avenue, New York, N.Y. 10016

Library of Congress Cataloging in Publication Data:

Leonard, George Burr, 1923-
The silent pulse.

Includes bibliographical references.
1.   Rhythm—Psychological aspects.   2.–Personality.
I.–Title.

BF475.L46 1978   152.3   78-18202

ISBN: 0-525-20450-4

Published simultaneously in Canada by
Clarke, Irwin & Company Limited, Toronto and Vancouver

10 9 8 7 6 5 4 3

For my aunt
EMILY LEONARD
in appreciation of
many years of devotion

CONTENTS

ACKNOWLEDGMENTS

Leo Litwak suggested the initial idea for this book, as he did for my last one; I thank him for his support and friendship. Susan Trott Mitchell provided invaluable advice and encouragement. Michael Murphy, as always, illuminated my explorations with his rare enthusiasm.

My thanks go to Carter Brandon for generous research assistance, to Saul-Paul Sirag for tutoring in quantum theory, and to Karl Pribram for his counsel on the brain and senses. I also want to thank John and Julia Poppy, Brendan O'Regan, Sterling Lord, and my aikido colleagues, Richard Heckler and Wendy Palmer.

I owe a great deal to the many participants in my workshop sessions, and particularly to my students in a special ongoing workshop at Aikido of Tamalpais in Mill Valley, California. Their wholehearted participation has contributed a great deal toward creating and validating the exercises presented in the Appendix of this book.

Finally, I want to express my deep gratitude to my editor, Bill Whitehead, for his broad understanding, his close attention, and his personal commitment to this work.

PREFACE
The Way of Rhythm and Harmony

The sperm cell swims with rhythmic strokes and joins the egg. Molecules of DNA dance together. Pulselike concentrations of fields interact, multiply, differentiate. A singular pattern emerges, something unique in the universe: a new being.

Remembering everything, the being passes through the various stages of earthly evolution, accompanied by the powerful drumbeat of the mother's heart. The being is shaken to the core by these pulsations, which promise purpose, wholeness, synchrony. Secure in this rhythm, the being's own heart takes form and begins an answering pulse.

As soon after birth as possible, the mother takes the baby in her arms and puts its head against her heart. The rhythm is still there, a reliable beat against which to measure the flow of growth and change. Later there will be other rhythms, other relationships. But some deep knowledge of those early moments remains, a reminder of the rhythm that sustains life and underlies all of existence.

This book is about that underlying rhythm. We begin, as Pythagoras did some 2,500 years ago, with music, in which is encoded the basic structure of the universe. From there we go to the hidden rhythms of human relationships, then deep into the body, where what seems most substantial eventually dissolves into patterns of pulsing waves, and the senses, which serve in marvelous ways to connect us to the rhythms of the world. This perspective leads us to recent research findings, from which we can draw a startling picture of human nature and human abilities: *At the heart of each of us, whatever our imperfections, there exists a silent pulse of perfect rhythm, a complex of wave forms and resonances, which is absolutely individual and unique, and yet which connects us to everything in the universe. The act of getting in touch with this pulse can transform our personal experience and in some way alter the world around us.*

Though I have drawn upon scientific theory and research in arriving at this point of view, especially upon certain implications of modern quantum physics, *The Silent Pulse* is by no means a formal scientific or philosophical treatise. Rather, it is a search, a journey of discovery involving personal experiences and stories about ordinary and extraordinary people as well as research information.

If what emerges from this search seems surprising or even fantastical, that is perhaps inevitable. For once you start seeing the material world in terms of rhythmic waves, your ideas about what is "real" and "true" will probably seem much less certain. In place of that certainty, I hope that you will entertain a new sense of the richness and endless possibilities of existence.

At the end of this book you will find an Appendix describing a number of exercises. Based on workshops I have given for more than 20,000 people over the past several years, these exercises are designed to lead you toward the experience of what I call perfect rhythm. The main body of the book is complete in itself. The Appendix is offered to those who wish to go a step further.

ONE
Vibrato

The man that hath no music in himself,
Nor is not mov'd with concord of sweet sounds,
Is fit for treasons, stratagems, and spoils;
The motions of his spirit are dull as night,
and his affections dark as Erebus:
Let no such man be trusted. . . .
    —*Lorenzo in* THE MERCHANT OF VENICE

There are human societies so simple and una-dorned as to possess no clothing other than the loin-cloth, no tool other than the stick or stone, no perma-nent dwelling place, no carving or other plastic art. But nowhere on this planet can you find a people without music and dance.

In New Guinea, a tribal chieftain coaxes hypnotic overtones from a Jew's harp made of a living, buzzing beetle. Half a world away, a band of BaMbuti Pygmies awakens the jungle with happy singing and clapping. High in the Himalayas, temple musicians shake the valleys with the awesome wail of twenty-foot-long horns, while priests in meditation produce three tones simultaneously (a lifetime's devotion) from a single voice. In Trinidad and in Central Park, West Indian musicians strike Calypso tunes out of the cut-off tops of used oil drums. There is no end to this global feast of sounds old and new, now being spread out electronically for all to sample.

Not so many years ago, I sat in a New York apartment and phoned friends in London and San Francisco, one after the other. In the background, during both calls, I could hear the Beatles' recently released *Abbey Road* playing—which is exactly what happened to have been playing in the next room there in New York. For a moment I had the feeling that I could go on phoning people on every continent and keep hearing the same music. A world in harmony!

But whether we make music in concord or diversity, in the jungle or the concert hall, we do make music. We make music because we are human. Seeing pyramids and cathedrals, factories and skyscrapers, cities and empires, we might be led to agree with the opinion that "man was born a builder," and that men will go on building until every meadow sprouts high-rise buildings and every river runs in its own concrete channel. This opinion finds no support in the anthropological record. The truth of the matter is, the human animal was born a singer and dancer. Rich and complex cultures existed for at least a million and a half years without pyramids and cathedrals, and we can perhaps conceive of a future without high-rises. But a humanity without music is not just inconceivable; it is impossible.

At the root of all power and motion, at the burning center of existence itself, there is music and rhythm, the play of patterned frequencies against the matrix of time. More than 2,500 years ago, the philosopher Py-

thagoras told his followers that a stone is frozen music,
an intuition fully validated by modern science; we now
know that every particle in the physical universe takes
its characteristics from the pitch and pattern and over-
tones of its particular frequencies, its singing. And the
same thing is true of all radiation, all forces great and
small, all information. Before we make music, music
makes us. The blessed gift of hearing serves as a channel through which we can be reminded of our deepest
origins. For music is a reflection in sound of the
world's structure, making explicit the rhythmic quality
in all things, which otherwise we might only deduce or
infer.

It's fitting, then, that we begin this exploration of
ourselves and of the world with music, and more specifically with a musical quality called *vibrato*. This pulsation that wells up within the sounded note can lead
us to what is most spontaneous and creative in human
life, and possibly even to deeper mysteries—to powers
of knowing and doing which we have lost or given
away during the epoch of civilization, and which perhaps we may now regain.

I have a friend who has a magnificent voice. Trained
as a classical pianist in his childhood, he turned from
music to the ministry, then to Civil Rights activism and
deeds of civil disobedience that gained him national
fame. His activities and concerns are of great interest
to me, but when he comes to my house, I steer him as
quickly as I can to the piano. He sits there for a moment—this square-jawed, full-chested man—grinning
at the piano as if it were a long lost friend. He strikes
a chord or two, then begins playing an accompani-
ment, meanwhile singing softly, in barely a whisper, of
the River Volga. His fine baritone starts to swell. He
swings into an irresistible rhythm and we find our-

selves clapping in time to the music. At this, his voice becomes truly enormous, with a full-bodied vibrato that comes all the way up from the floor. Small objects in the room begin to vibrate—candlesticks, glasses on the coffee table, a vase on the mantlepiece. I have the feeling that the sound of his voice literally *fills* the room. Nothing is empty or lacking. Every corner is full, every crack, every crevice. Inanimate objects are infused with life—the chairs, the lamps, the paintings on the wall. The sound that comes from his mouth seems to come from everywhere. "Inside" and "outside" lose their meaning. And I am filled with this music down to the marrow of my bones.

Without vibrato, my friend's voice would still be a remarkable phenomenon. With vibrato, it is for me an expression of the *élan vital,* the elusive yet ubiquitous force that imbues the inanimate material world with life and intentionality. *Webster's International* defines vibrato as "a slightly tremulous or pulsating effect (but not a tremulo) for adding warmth and beauty to the tone or for expressing changes in emotional intensity. It consists of slight and rapid variations in the pitch of the tone being produced and is most effective in comparatively long tones."

Where tremulo is a forced variation of the tone itself, vibrato comes from *within* the tone, filling it out. Vibrato is intrinsic, spontaneous, and can be profoundly erotic. Sometimes, when the slow movement of Bach's *Fourth Brandenburg Concerto* begins and all the woodwinds and strings start trembling and pulsating together, I glance at the people around me, experiencing a certain wonder that such an intimate sound should be so casually produced in public.

The Victorians, as a matter of fact, considered vibrato to be voluptuous and even sinful, and singers of that day took pains to minimize the natural vibratory quality in their voices. The young music critic, George Bernard Shaw, in one of his more prudish moments, inveighed against the late Victorian comeback of vibrato which, he wrote, was sweeping "through Europe like the influenza."

The "dangerously sensual" quality of vibrato might be explained by the fact that its rate of vibration, about seven pulsations a second,* precisely matches the theta-wave state of the brain. This state is associated with the twilight zone between waking and sleeping, in which the customary censorship of the conscious mind is absent. Vivid hypnogogic images (from the Greek *hypnos*, "sleep," plus *agogos*, "leading") pop into the mind spontaneously, as if from nowhere. The waking dreamer, in fact, sometimes seems to have access to all the wells of memory and creation, perhaps to some sort of group consciousness. Elmer and Alyce Green of the Menninger Foundation have reported a number of extraordinary psychic experiences during the theta state. They also point to the classic examples of creative artists and scientists who have received inspiration through this kind of imagery—William Blake, John Milton, Samuel Coleridge, Robert Louis Stevenson, Jean Cocteau—and to Poincaré's well-known description of a vision he beheld as he lay in bed awaiting sleep: mathematical ideas dancing in the clouds before him, and colliding and combining into what he recognized as the first set of Fuchsian functions, the solution to a problem he had long been struggling to solve.†

* In the 1920s and '30s, Dr. Harold Seashore measured the vibrato rates of several famous singers. Caruso sang at 7.1 cycles a second, Galli-Curci at 7.4, Martinelli at 6.8. The rate changes with the demands of the situation, speeding up against a full-bodied orchestra, slowing during moments of tenderness.

† Elmer and Alyce Green, *Beyond Biofeedback* (New York, 1977), p. 125.

A powerful musical vibrato might have the effect of capturing the rhythm in our own brains, thus creating the condition of reverie in which mysteries are revealed. Listening to music, especially the slow, sustained passages that allow the full development of vibrato, I have sometimes been "drawn into a spell." Could it be that the "enchantment" of music has a rhythmic, physiological basis?* In any case, vibrato represents the spontaneous and the emotional rather than the rigid and the rational—a threat to any repressive age, a joy and a beacon to the creative.

When we hear an orchestra, we think of the drum and other percussion instruments as providing the rhythm. Actually, every sound that comes from every instrument is rhythm. We can make out the separate beats of a drum, but should those beats come faster and faster, eventually they would blur into what we would perceive as a musical tone. The slowest rate of vibration that we recognize as "music" might be the lowest note on the piano, which pulses at 27.5 beats a second. The highest note on the piano pulses 4,186 times a second. The frequency range of the human ear is even greater—normally from around 20 to 20,000 cycles. As dog trainers know, some creatures can perceive still higher sounds.

When my friend hits a high E in one of his Russian songs, his vocal chords are vibrating approximately 659 times a second. But that, the basic pitch of the

* One recent study has shown that a standing wave produced by the heart-aorta system produces a small seven-cycle-a-second oscillation in the human skeleton. See Itzhak Bentov, "Micromotion of the Body as a Factor in the Development of the Nervous System," in Lee Sannella, M.D., *Kundalini—Psychosis or Transcendence?* (San Francisco, 1976), pp. 71–95.

tone, is only a part of what we call music. The human voice, like all musical instruments, produces over-tones, vibrations of higher pitch, which create the rich-ness of tone we find pleasing. In fact, the sounding of a single note on a single instrument, say a violin, is a matter of incredible complexity, as Guy Murchie ex-plains in his encyclopedic *Music of the Spheres:*

> Both volume and pitch interrelate with timbre or tone quality, which is a kind of sum product of all the details through which vibratory energy dis-tributes itself—as, for instance, the stick-slip-stick-slip-stick friction waves of the violin bow upon a string that blends the fundamental tone (voiced by resined horsehair on sheepgut) with its harmonics, flowing into the oaken bridge and headlong down the spruce soundpost, spreading thence impetuously across the chamber of the body and through the cells of the surrounding planewood, beech or pine, shaped by interrelated contours, graceful *f*-holes, invisible nodes, and on outward into the air in all directions at better than 1,000 feet a second, to reverberate anew to the shape of the hall or the natural resonance of surrounding trees and buildings. Thus the strains you hear are literally the living compend of a hundred factors from the hand of the musician to the pitch of the balcony—even to the tensions in your own ear and brain . . . expressing a sort of tuned aspect of the entire local world.*

When the single note from a single violin is com-bined with notes from other violins and other instru-ments and voices, and when all these sounds begin to

* Guy Murchie, *Music of the Spheres* (Cambridge, Mass., 1961), pp. 398–99.

change with time to create melody and counterpoint, the result is an intricate, interwoven complexity of rhythms that stands beyond solution in terms of relationships of individual parts. The complexities resolve themselves esthetically, as a whole. The "solution" is a single, vivid, unified emotion.

The way that music works, as we'll see in this book, is also the way the world of objects and events work, for it is all vibration. For example, if you should glance for only a second at the yellow wing of a butterfly, the dye molecules in the retinas of your eyes will vibrate approximately 500 trillion times—more waves in that second than all the ocean waves that have beat on the shores of this planet for the past 10 million years. Were the butterfly blue or purple, the number of waves would increase, since those colors vibrate faster. With X rays instead of light, the rate of vibration would speed up a thousand times, with gamma rays a million. (The vibratory rates of the subatomic particles that make up ordinary matter are incredibly higher, while the waves at the heart of the atom's nucleus vibrate at a rate that would strain our imagination, as will be seen in Chapter 4.)

The rates of vibration of all radiated energy—including radio waves, heat, light, X rays, and so on—can be arranged in order. The resulting electromagnetic spectrum turns out to have more than seventy octaves, with visible light being only one of these octaves. As in music, all the "tones" in this spectrum have their own harmonic overtones, and there are certain similarities that appear at octave intervals. Many of the most basic discoveries of the scientific world, in fact, have simply confirmed the musical nature of the world. For instance, the Periodic Table of Elements, in which all the chemical elements are listed in order of atomic weight, breaks down into seven octaves, with properties that tend to repeat, as in musical octaves. No wonder so many scientists and mathematicians are also musicians.

Were Pythagoras alive today, he would be delighted but not surprised to learn that the rigors of the scientific method have succeeded in uncovering a musical, mystical inner order of nature more beautiful than anyone other than a Pythagorean might have anticipated. And Johannes Kepler, the seventeenth-century astronomer who worked out the laws of planetary motion, would feel a certain satisfaction about a recent experiment by two Yale University professors.

Kepler believed that each of the planets was alive and inhabited by its own guardian angel, who alone could hear its music. He went so far as to work out each planet's "song" in terms of its orbit around the sun. Willie Ruff, an assistant professor of music, and John Rodgers, a professor of geology, took Kepler's laws and musical notations and applied them to the motion of the planets as projected over a 100-year period starting December 31, 1976. They fed this information into a computer connected to a music synthesizer. What emerged was a thirty-minute tape representing 100 years of planetary motion.

"I wanted to figure out whether Kepler was the screwball he was said to be," Professor Ruff explained, "or whether he was actually on the right track. I wanted to know, 'Will it come out? Will it be what Kepler promised?'—that is, a continuous song for several voices."

It came out as a spectacular if somewhat dizzying piece of music, with Mercury, the fastest-moving planet, singing the shrill ascending and descending slide of a piccolo, and Jupiter, the slowest, sounding a deep, powerful rumble.

"Venus changes from a major to a minor sixth, and earth makes a marvelous minor second," Professor Rodgers said. "In fact, it's just as Kepler said."

Rhythm, resonance, and harmony order the heavens. And sound, through which music is expressed, has a particularly significant quality in our earthly life, the quality of all-aroundness, of the here-and-now. We may get most of our specific information through vision, but it is hearing—as will be seen in the next chapter—that connects us most intimately with the matrix of our existence. In terms of learning ability, problems with the hearing mechanism are almost always more disabling than problems with vision. Music and dance, the blending of rhythmic sound with movement, lubricate all learning, and it is tragic that these subjects are considered anything less than basic in our schools.

This is not to say that every child can be made a consummate musician. Musical abilities can be cultivated and the appreciation of music learned, but musical genius is a gift. A Mozart emerges now and then to affirm the connection that is possible between individual consciousness and ultimate form, a connection that will be one of the main subjects of this book. The musical realm is essentially amoral; men and women who are without moral sentiment can be possessed of an exquisite musical sense. The absolute lack of any feeling for music, however, does seem to indicate that something is awry, for to be without music entirely, a rare malady indeed, is to be out of tune with the world.

Just as music reminds us of the nature of existence, the vibrato that wells up out of music reminds us of the spontaneous powers existence offers us. The trembling hand of the string player, the trembling lips or throat of the wind player or singer are emblems of that power. Not long ago I read a newspaper account about

a group of ghetto children taken to their first symphony concert. One boy of eleven, asked for his reaction after the concert, said that the musicians had been
"scared." He knew this was so because, "I saw their
hands shaking." The other children agreed: "They
were scared, all right."

And who is to say these first-time concertgoers were
mistaken? Involved in the expression of the beauty
and power that is music, listeners and musicians alike
must feel at least a twinge of awe, perhaps of fear. In
the poet Rilke's words:

> . . . Beauty's nothing
> But beginning of Terror we're still just able to bear,
> And why we adore it so is because it serenely
> Disdains to destroy us.

The particular arrangement of sound that we call
music has to do, finally, with relationship itself, with
the way things are similar and the way they are different, the way things grow and decay and the way new
forms evolve from growth and decay. The deep structure of music is the same as the deep structure of
everything else. When you impose ornamentation
onto music from the outside, it will turn out as cheap
and false. In good music, the deep structure creates
the ornamentation—trills, grace notes, even vibrato.
Vibrato comes not from the surface but from the
depths. Vibrato is inevitable simply because, at the
heart of it, the world vibrates.

So let us examine existence as it is, as music and
rhythm, following the beat of life wherever it may lead
us. This journey begins on the familiar ground of
human relationships, then plunges down through the
depths of the body, where nothing is quite as it has
seemed. Later, we may enter forbidden terrain and

encounter strange vistas that might offend some conventional views of what is real. I assure you I have no desire whatever to shock or offend. To the contrary, I am keenly aware of what Plato pointed out in ancient times: the best anyone can hope to do is remind you of what you already know. My best hope, then, is not for a journey of discovery, but for one of remembrance.

TWO
The Rhythms
of Relationship

In 1665 the Dutch scientist Christian Huygens noticed that two pendulum clocks, mounted side by side on a wall, would swing together in precise rhythm. They would hold their mutual beat, in fact, far beyond their capacity to be matched in mechanical accuracy. It was as if they "wanted" to keep the same time. Huygens assumed a sort of "sympathy" between the two of them, and conducted experiments to learn just how the interaction took place—whether they were linked through the air or through the wall on which they hung. From his investigations (the pendulums, he found, were synchronized by a slight impulse through the wall) came the first explanation of what scientists were to call "mutual phase-locking of two oscillators" or simply "entrainment."

This phenomenon, as it turns out, is universal. Whenever two or more oscillators in the same field are pulsing at *nearly* the same time, they tend to "lock in" so that they are pulsing at *exactly* the same time. The reason, simply stated, is that nature seeks the most efficient energy state, and it takes less energy to pulse in cooperation than in opposition. Entrainment is so ubiquitous, in fact, that, as with the air we breathe, we hardly notice it. Yet it offers dramatic witness of the tendency toward perfect rhythm that we discover whenever we examine the roots of our existence.

To get the feel of entrainment, you might try playing with the "vertical" and "horizontal" knobs of an old television set. Every set contains horizontal and vertical oscillators that position the scanning electronic dot that forms the picture. These oscillators must match the signal coming from the station very precisely; otherwise the picture will move sideways or vertically. When you turn the knobs, you're adjusting the frequency of your set's oscillators to match the frequency of the station's oscillators. Fortunately, you don't have to create a perfect match. When the frequencies come close to one another, they suddenly lock, as if they "want" to pulse together.

You might note the sense of satisfaction and "rightness" that comes when the picture locks securely to its frame. But what if the set is defective, and the picture won't hold? There's probably a sense of frustration, even rage, especially when the mechanism is only slightly off. You adjust the vertical hold. It seems perfectly all right. You sit down to watch your favorite program. . . . Just as you get comfortable, the picture slips out of its frame. Nothing seems more perverse or is more infuriating than a slight flaw. Television engineers have worked hard to solve this problem, and the newer sets have effective automatic hold mechanisms.

Living things are like television sets in that they contain oscillators. In fact, we might say that living things *are* oscillators; that is, they pulse or change rhythmically. The simplest single-celled organism oscillates to a number of different frequencies, at the

14

atomic, molecular, subcellular, and cellular levels; microscopic movies of these organisms are striking for the ceaseless, rhythmic pulsation that is revealed. In an organism as complex as a human being, the frequencies of oscillation and the interactions between those frequencies are multitudinous. Most of our internal systems operate in rhythmic relationship; many of the component parts must be synchronized. There is an electrifying moment in the film *The Incredible Machine* in which two individual muscle cells from the heart are seen through a microscope. Each is pulsing with its own separate rhythm. Then they move closer together. Even before they touch, there is a sudden shift in the rhythm, and they are pulsing together, perfectly synchronized.

Just as our internal rhythms are locked on "hold" with one another, they are also entrained with the outside world. Our physical and mental states change in rhythm with the seasonal swing of the earth and the sun, with the tides, with the day-night cycle, and perhaps even with cosmic rhythms that present-day science hasn't yet isolated and defined. When these rhythms are forced out of phase, disease is likely and dis-ease is inevitable.

How about human relationships? Is there some sort of "hold" mechanism involved when you talk to someone, something that goes deeper than the verbal content? According to the pioneering work of Dr. William S. Condon of Boston University School of Medicine, this is indeed the case. Dr. Condon's discoveries about the subtle, largely unseen conversational dance, in fact, tend to upset our usual notions about the interaction between speaker and listener, and suggest a new way of looking at all relationships.

To investigate what the naked eye can't see, Condon has devised a method he calls microanalysis. He takes moving pictures of, say, a conversation between two people, then displays the film on a time-motion analyzer. He numbers the individual frames and runs them manually again and again, a thousand times if need be, until every micromovement is noted. Words and syllables of the speaker's voice are also broken down into tiny units. For example, the word "ask," on one film (shot at 48 frames per second), is seen to last one-fifth of a second. During this period, however, there are four smaller units: /ae/, lasting 3/48 second; /E/, lasting 2/48 second; /S/, lasting 3/48 second; and /k/, lasting 2/48 second.

As it turns out, the micromovements of the speaker's body are precisely synchronized with the microunits of his speech. During the 3/48-second duration of /ae/ on this film, for example, the speaker's head moves left and up slightly while the eyes hold still. The mouth closes and comes forward. The four fingers begin to flex and the right shoulder rotates slightly inward. All these movements continue as a single, flowing unit until the end of the sound. Then, with the beginning of the second sound in the word "ask," which is /E/, another distinct cluster of movements begins, and so on.

These tiny units of movement occur *within* longer units, ranging up to the conventional, visible speaker's gestures with which we're familiar. All the movements, short and long, follow the same beat, linking syllables, words, and phrases to the dance of the body in precise rhythm. Here it's important that we not think of the pattern of movement, of beats within beats, as merely the ornamentation of our talk. "Rhythm or time must be seen," Condon explains, "as a fundamental aspect of the organization of behavior and not as something added to it."

Dr. Condon had first started using film analysis in

the early 1960s for a quite different purpose: to study
the relation of conventional gestures to verbal material. But he noticed that something else was going on
beneath the more obvious gestures. So he began his
painstaking analysis of micromovements. He spent a
year and a half looking at one 4 1/2-second conversational sequence between two people, wearing out 130
copies of the film in the process. (He figures a film can
be run 100,000 times before wearing out.)

His discovery of the entrainment between the
speaker's words and movements, Condon told me,
came as a surprise to him. He was still more surprised
when he recognized what he called "a very startling
phenomenon." Dr. Condon's sense of wonder at this
second discovery comes across even in the formal language of a scientific paper:

> *Listeners* were observed to move in precise
> shared synchrony with the speaker's speech. This
> appears to be a form of entrainment since there
> is no discernible lag even at 1/48 second. . . . It
> also appears to be a universal characteristic of
> human communication, and perhaps character-
> izes much of animal behavior in general. Com-
> munication is thus like a dance, with everyone
> engaged in intricate and shared movements
> across many subtle dimensions, yet all strangely
> oblivious that they are doing so. Even total stran-
> gers will display this synchronization. Such syn-
> chronization appears to occur continuously if the
> interactants remain attentive and involved. . . .
> A listener usually does not move as much as a
> speaker. There will be moments when he remains
> quite still. The specific body parts and their direc-
> tion of movement often differ from those of the
> speaker. But if the listener moves, even when

reaching for a pack of cigarettes, etc., his movements will tend to be synchronous with the articulatory structure of the speaker's speech.*

If Condon is right, those of us accustomed to thinking in the usual action-reaction, stimulus-response way about human behavior will have to modify our thinking. At the most fundamental level, the listener is not *reacting* or *responding* to the speaker. The listener is in a sense *part of, one with* the speaker. This becomes startlingly clear when there is a silence in one of the conversations filmed by Dr. Condon. *At the precise 1/48 second the speaker resumes talking, the listener begins his or her series of synchronized movements.* Exactly how this is possible cannot be explained in terms of conventional psychology or physical theory.

Go back to the day of birth, and the mystery deepens. Condon's film studies show that a normal newborn infant, like a normal adult, moves synchronously with the pattern of the mother's voice. The infant obviously doesn't understand the meaning of the words, which could be English or Chinese or Swahili; but it is now becoming clear, from a number of studies, that literal content is only a part, probably a minor part, of human communication. The infant is born *already connected* with the fundamental rhythms that hold us all together.

"I see the world as altogether unified," Dr. Condon told me. "The brain of the baby is also unified and interfaced with an ongoing rhythmic unity that represents its world. Infants are born with the order of the world already in them."

There are exceptions, children who cannot prop-

* William S. Condon, "Multiple Response to Sound in Dysfunctional Children," *Journal of Autism and Childhood Schizophrenia* 5:1 (1975), p. 43.

erly "hold" the world, and in the tragic history of those we call "dysfunctional" we can read the crucial importance of rhythmic entrainment. Condon filmed and analyzed the responses of twenty-five children ranging in age from eight days to twelve years, most of whom were classified as "autistic-like" or "reading problem and slow learner," and made another totally unexpected discovery. These children's micromovements entrained with speech in the normal way while it was going on. But then there was an echo. After an interval of up to a second, their bodies moved again, as if in delayed response to the earlier sound. Some of the children seemed to *hear* the nonexistent delayed sound; others did not. In every case, however, the hearing delay caused distinct bodily movements. The children responded even to their own voices, as if they, too, were being repeated to them after a slight delay.

According to Dr. Condon, "One of the most marked features of the response to delayed sound is the seeming lack of control of these children over their own bodies. They seem as if jerked this way and that by the sound. Most of them tend to move more at the delayed time than when the actual sound occurs."

Such a dissonant world is perhaps unimaginable to one who has not suffered it, and indescribable to one who has. No wonder the severely autistic child simply withdraws from that nightmare. Our ability to *have* a world depends upon our ability to entrain with it.

In this light, much that previously was unclear takes on new meaning. The newborn baby's ability to entrain with his or her mother's speech provides a mechanism that helps explain the bonding process which more and more psychologists are seeing as crucial to normal development. It also helps us understand the value of natural childbirth in pleasant, quiet surround-

19

ings, and to realize why isolated nurseries full of crying babies are such terrible places.

Semantic meaning gives us directions, but the pulse within the sound sustains our life. In focusing our attention on language as a symbol system, in fact, we have tended to neglect the role of rhythm in the creation of speech and the evolution of languages. In the words of linguist R. H. Stetson, "The rhythm is certainly one of the most fundamental characteristics of the utterance of a language, and is often most difficult for a foreigner to acquire. . . . There can be little question that a profound change in rhythm underlies the extensive alterations during the transition from Latin to French. *A new rhythm has produced a new language,* reshaping the ancient words, eliminating syllables and shifting the stress to an alteration from syllable to syllable. . . . Such a change of rhythm and stress is apparent in the Hebrew over against the Arabic and in the Czech over against the Lithuanian, and in the French over against the Italian."*

I have provided italics in Dr. Stetson's quotation to underline a key point: A culture may be said to *have* a language, but a culture *is* a rhythm. This point impresses itself on us here in America, where two cultures, black and white, with the same ostensible language, have had serious problems in communicating. We are beginning to realize that Black English—rich, specific, highly expressive—is actually different from White English. But grammar, syntax, and vocabulary can be mastered. The deeper problem lies in the rhythmic differences.

Culture theorist Edward T. Hall, using Condon's

---

* R. H. Stetson, *Motor Phonetics: A Study of Speech Movements in Action* (Amsterdam, 1951), pp. 124–25.

methods as well as his own, has found what he calls "great differences" between working-class blacks and a wide range of whites in such nonverbal behaviors as kinesics (body movements), proxemics (use of personal space), and rhythm in general. Hall argues that these nonverbal modes of communication are "interwoven with the fabric of the personality and into society itself, even rooted in how one experiences oneself as a man or woman."*

This is a tricky business. Veiled prejudice can lie behind admiring statements about ethnic groups that have "got rhythm." The fact is, every ethnic group has got rhythm—but the rhythms are different from one another. And members of majority ethnic groups have often acted as if anything different has got to be inferior.

During the period of greatest racial tension in the late 1960s, I conducted a number of interracial marathon confrontations with Dr. Price Cobbs, a black psychiatrist. The format was simple. We brought together groups averaging around twenty participants, mixed as to race and sex. We met for an introductory session of three or four hours on Friday night. Then, at noon Saturday, we set ourselves in for twenty-four hours of talk, with the hope that we would break through the invisible barriers that separated race from race. No physical violence was permitted but, as far as words were concerned, no holds were barred.

The results were consistent. For hours on end, torrents of verbal abuse would pour out—resentment, fear, hurt, rage. Sometime before dawn, there would come an interlude when the situation would seem absolutely hopeless. But after sunrise, beyond all reason,

* Edward T. Hall, *Beyond Culture* (Garden City, N.Y., 1976), p. 71.

21

the current would start to shift. Resentment would turn to understanding. Tears of anger would become tears of sympathy and awareness of the mutual bondage that had grown out of racism. The barriers were shattered.

Why were these sessions successful? Dr. Cobbs and I credited freedom of expression on an emotional, here-and-now basis. We noted the role of sleeplessness in breaking down preconceptions. We appreciated the classical dynamics of catharsis. These factors undoubtedly did play their part. But it was not until I came across the work of Condon and Hall that I recognized the interracial marathons as crash courses in rhythmic entrainment.

Hall's studies explain the trouble that blacks and whites have in becoming synchronized with each other. This difficulty expressed itself in our groups by a pervasive weariness that seemed to come over us at the very beginning of the marathon. It was as if words didn't have their usual force and effect, as if all of us on both sides of the racial barrier were fighting our way through heavy mud. Perhaps the long hours of talk, the sleeplessness, the catharsis, were needed for us to get our conversational rhythms together, to begin the process of entrainment that is the essence of human communication. There's no question but that the rhythm, the *feel* of the talk (preserved on tape recordings) is quite different near the end of sessions than during the early hours.

There is a section on one of the tapes that is particularly revealing. It comes around the middle of a marathon, and consists of practically everyone in the group talking at once, cursing, shouting, stamping their feet. Playing this section again and again, I began to hear it, not as words in the usual sense, but as music—a mad, Wagnerian crescendo and diminuendo, having its own internal rhythm, and even a rising and falling pitch. Near the end of the section, some of the shouts and curses began turning into laughter. Then a strange thing happens: The entire group suddenly stops, then begins again, then stops, then begins again more quietly—all in perfect rhythm. After this, the encounter resumes with a new tone of tenderness and ease. It's as if the pendulums of understanding are swinging together, the heart cells beating as one.

Dr. Paul Byers of Columbia University, a researcher in nonverbal communication, has analyzed filmed interactions of Americans, the Maring of New Guinea, the Netsilik Eskimo, and the Kung Bushmen of Africa. In every case, he has found that rhythm-sharing is present. Byers uncovered one bizarre example of the use of rhythm in films of the Yanamamo, who live on the upper Orinoco in South America and are known as The Fierce People. Whenever two villages come together for a feast, the visitors act toward their hosts in an extremely aggressive and threatening manner. The chiefs of the two villages confront each other in the center of the village and engage in a fierce shouting match. But things are not what they seem on the surface, as Byers found out when he used machines to analyze the precise rhythm of the interchange:

Each man shouts three syllables which begin exactly two-tenths of a second apart. But one man starts exactly one-tenth of a second after the other man. The next burst begins exactly two-tenths of a second after the stress peak of the second man's last (third) syllable. The result sounds like an angry shouting match but, in fact, represents an incredibly tightly synchronized talk-dance—a precise and phase-locked interaction. The shouted content suggests angry aggression, but the process has the biological effect of making the interactants feel good about what they are doing—and about each other. This precision is possible because the procedure synchronizes the brain waves of the two men and the brain waves, which pace the underlying motor behavior, are almost mechanically precise.*

Byers points out that a variety of bodily processes can become synchronized through close interaction. "Singing, rowing, and sometimes marching will synchronize breathing. Synchronized heartbeats have been reported between psychiatrist and patient. Female college roommates sometimes find their menstrual cycles synchronized."

"The more you move in rhythm with someone," Dr. Condon told me, "the closer you become with that person." This is true not only with human beings but with all living creatures. Condon suspects that entrainment is involved in the movements and sounds that precede mating in most animals and insects. Some of these courtship dances are incredibly complex. One species of ruff (a common European sandpiper) uses at least twenty-two separate visual displays in mating, with males of different ranks identified by the subsets of signals they use. Certain species of grasshoppers go through even more complicated rituals. In every case, sounds and movements trigger hormonal secretions that ensure success in mating and procreation. The creation of life is indeed a rhythmic process. Members of the human species may override the natural rhythms of sex (rape is the ultimate disrhythmic

* Paul Byers, "A Personal View of Nonverbal Communication," *Theory into Practice: Journal of the School of Education* (Ohio State University, June 1977).

act), but there is no true sexuality without rhythm shared.

In music, the miracle of entrainment is made explicit. The performer's every gesture, every micromovement, must be perfectly entrained with the pulse of the music, or else the performance falls apart. Watch the members of a chamber group—how they move as one, become as one, a single field. We have become accustomed to such miracles: the extraordinary faculty of jazz musicians to "predict" precise pitch and pattern during improvisation, the simultaneous sweep of sixty bows in a symphony orchestra. The miracle springs not so much from individual virtuosity and sectional pyrotechnics as from the ability of a large group of human beings (hundreds in oratorios) to sense, feel, and move as one. The conductor of the symphony can be viewed as an absolute ruler, but he rules by playing with—that is, becoming entrained with—his orchestra. And entrainment, again, involves no stimulus-response, no action-reaction. The conductor can be truly despotic in creating the necessary and sufficient field of play. But when the music begins, if the performance is to be a great one, he and all his musicians are equally enthralled.

My own life has held music at its core. I began my musical studies at nine, and twice came close to taking up music as a career. But the years have passed and I have accepted the less strenuous joys of an amateur jazz pianist and occasional composer of songs for modest theatrical productions. It is precisely my lack of polish as a pianist that allows me to examine the question of entrainment in terms of two distinct musical states. In the first state, I am improvising consciously, forcing the music rather than becoming one with it. This improvisation, if we can call it that, is a matter of editing rather than creating; I am simply picking out riffs from past improvisations and placing them in some order that satisfies the requirements of the melodic line and the harmonic progression. My movements in this state are studied and awkward. The music is stilted and joyless.

Then, if all goes well, there is a transformation of state. It is not a gradual change but a sudden shift, marked for me by an unmistakable physical sign: whatever the room temperature, I feel a flush of perspiration on my forehead. At this moment, the improvisation becomes effortless. I take no thought of what I'm going to play; I simply allow my consciousness to stay in the time and place from which the music emerges. Within my technical limitations, I feel that I can "do anything." This entirely unreasoned music, it turns out, contains a far more satisfying formal structure than that produced by means of conscious thought.

As for my physical movements, they are, in the second state, entirely spontaneous and joyful, *part* of the music. Such movements can appear strange and even grotesque, especially in jazz and rock music. A famous jazz pianist once told me that he asked his wife not to come to his most important concerts and recording sessions. He well knew that at his creative best his movements and gestures would become excessive. He was embarrassed to have the one he cared for most see him in such a condition.

The point here is that every creative act, every act of interaction, has its own characteristic and appropriate movements and gestures, which can be suppressed only at the expense of what is spontaneous and ultimately nourishing. These movements and gestures are shaped to some extent by cultural norms. The dance, song, and shamanistic possession of the Natsilik Eskimo involve different rhythms and different bodily movements from those of the Nigerian Yoruba.

But in both there is entrainment between sound and gesture, between participants, and, if you are willing to entertain such an idea, between self and cosmos. And if the culture should attempt to break the expression of this rhythm of relationship, it is the culture that eventually must yield.

For its own good reasons, the West has attempted for some centuries to suppress the "savage" expressiveness which asserts itself most vividly in possession and trance—or, at the least, to give it no positive value. Recent decades, however, have seen what Freud would call "the return of the repressed," in cathartic group practices, in charismatic religions, and, perhaps most strikingly, in the various musical forms derived from jazz: African rhythm and expressiveness married to Western musical form, the most significant cultural invention, with the possible exception of mass production, to come out of America. Let us witness soul singer James Brown on the stage at the Apollo Theater in Harlem—strangely anti-American yet totally American:

> Thus admonishing and exhorting us, he would caper, invertebrate, strutting, slithering, pirouetting, improvising a febrile ballet of black pride in India rubber; and bawling, bawling, bawling, driven to a Cyprian frenzy under the hypnosis of drums in league with a choir of keening saxophones underscored by an immense and brooding bass fiddle.
>
> And now uncontrollable passion assails him. Lying on one side, he stretches full length on the floor, supporting his weight with one hand while the other engages in unsupportable gropings. In simulated ecstasy he cries aloud: 'Shall I scream?' We are all suffering with him. 'Yes! Scream, soul,

scream!' we respond, thereby releasing him, and freeing ourselves. His need was ours; ours, his. We have achieved communication and understood each other.*

Entrainment with a high emotional content. No wonder we fear its power: The fire-and-brimstone preacher bombarding us with his overload of guilt and shame. The jingoist orator exhorting us to the rigid, rectangular rhythm of the military march. The psychedelic pulse of Hitlerian proclamation and the answering volleys of *"Sieg Heil!"*

Any successful public speech involves entrainment, but there are generally many in the audience who are thinking of something else. (A famous nonstatistic states that 82 percent of all people listening to sermons are thinking about sex.) During fifteen years of public speaking, I have experienced only a few of those awesome moments when I and the entire audience have begun to move and breathe as a single organism. Once, about midway through a talk to a large luncheon group in Los Angeles, a strange sort of silence overtook the audience. There were 1,200 people in attendance; tables on three levels stretched off into the distance. Yet the room seemed to get smaller and smaller as I spoke, and the gathering seemed to become almost unbearably intimate. I went on with my prepared speech; it was as if there were no way I could stop, or alter the rhythm. I felt myself and the audience drawn together as if by some powerful magnetic field. The experience was both deeply seductive and terrifying. I had no idea how it would end. Then, without warning, a woman about three tables back sobbed

* Frank Hercules, "To Live in Harlem," *National Geographic* (February 1977), pp. 188–93.

28

aloud, and the spell was broken. People once more began shifting in their seats, there were a few muffled coughs, and the speech continued on familiar ground, entrained in a more comfortable and customary manner.

There is no way we can escape it. Rhythmic entrainment, like all powerful processes, can be used for good or ill, to gain power or to withhold it from others. Rhythm can be captured or confused or broken. But only for so long. Ultimately, entrainment stands beyond manipulation, for it is the stuff of life itself, echoing the essential connectedness that defines existence. Different cultures speak and move in different specific rhythms, but the process of rhythmic connectedness is the same for all cultures. The autistic child is beset by contradictory rhythms, but at the heart of this life, as with all lives, there is the pulse of perfect rhythm.

Condon's painstaking studies of entrainment illuminate the extraordinary in the commonplace. Discussing dinner plans with a friend, you are dancing with precision and consummate skill. Talking casually about the weather with a stranger, you are joining two universes in ways previously unknown. Certain efficiency experts and others who take pride in the rigor of their relationships might denigrate small talk as a waste of time, and urge us to get right down to the point. What they fail to realize is that the function of small talk is entrainment, and entrainment, however accomplished, is the basis of all verbal interaction. More than that, it is our most accessible entry into the dance of the universe, and, as T. S. Eliot reminds us, "There is only the dance."

THREE
Flesh, Spirit, and Emptiness

The subtle dance of the body joins us to the world. But what is this body? Of what is it made?

The distinction has always seemed clear enough: Flesh is that chunky, solid stuff that is heir to ills, desires, ultimate decay. Spirit provides a contrast, having no odor, taste, or feel, no appetites. Flesh is substantial and vulnerable. Spirit is elusive and everlasting. Seeing the two in opposition, philosophers and religious leaders have had no hesitation in taking the spirit's side. For Plato, the world of matter and flesh was only a pale copy of a higher realm of Ideal Forms. Manichaeism, an early Persian religion that profoundly influenced St. Augustine, saw the whole universe as divided into two kingdoms. God ruled the kingdom of spirit, while Satan held sway over the kingdom of matter. The human body, being a gift of Satan, was totally evil. Thomas à Kempis referred to the body as "that dung heap." St. Francis vacillated between denigrating the body and reluctantly cooperating with what he called "Brother Body" or "Brother Ass."

The body, in short, has seemed hardly worthy of serious religious or philosophical consideration, being too stubborn, too obvious, too *solid*. Even the least philosophical of us has sometimes wished, with Hamlet, that "this too too solid flesh would melt,/ Thaw and resolve itself into a dew."

But is the body really solid? Our customary perceptions tell us that the skin is relatively smooth and inoffensive, while the flesh beneath is something terrible to behold: an oozing mass of sinews and tubes and *meat*. Through a microscope, however, the skin on a fingertip is a series of mountain ranges on a desolate moonscape, a crusting, flaking surface pockmarked with holes. A section of muscle tissue, on the other hand, appears as an elegant latticework, a thing of symmetry and beauty. Whether skin, flesh, or bone, however, this body of ours seems to be entirely solid and substantial.

We can penetrate more deeply. The electron-scanning microscope, with the power to magnify several thousand times, takes us down into a realm that has the look of the sea about it. Now the pores of the skin open like ocean caves, and we have to be told that the submarine creatures clinging to the convoluted walls are nothing more than ordinary bacteria.

Moving through the body with newborn perceptions, we observe a sea serpent lying on a giant walrus —actually a bundle of nerve fibers winding its way across segmented muscle fibers. Bulbous glial cells in the brain have the look of kelp rising from the floor of the sea. Ciliated cells in the fallopian tube appear as seaweed waving rhythmically in an undersea current. And now we see thousands of tadpoles swimming past, directly against the current, sperm cells engaged in a contest with unfavorable odds; the prize is a new life.

31

There is something dreamlike about these images. Auditory cells in the inner ear are like sea anemones in a tidal pool. And in the semicircular canals are giant boulders that tumble hither and thither on matted seaweed. The boulders are calcium crystals that move when the head moves; sensing this movement, the seaweed (receptor cells) signals the brain as to the head's position.*

Some Captain Cousteau of this silent world could spend a lifetime in the study of blood cells alone, which come in almost every conceivable guise. Most red cells are disks with depressions in the center. But some are like spiny balls or sea urchins. Others are shaped like bells or Mexican hats or Chinese spoons. And you could create a whole marine zoo of these creatures, not only sea horses, but sea hippos, rhinos, long-tailed rodents, lizards, and storks.

In the kingdom of corpuscles, there is transfiguration and there is *samsara*, the endless round of birth and death. Every passing second, some 2 1/2 million red cells are born; every second, the same number die. The typical cell lives about 110 days, then becomes tired and decrepit. There are no lingering deaths here, for when a cell loses its vital force, it somehow attracts the attention of a macrophage. This large scavenger cell, as formless and inexorable as The Blob of science fiction, approaches the doomed red cell, opens its huge round mouth, swallows, and digests it.†

The electron microscope allows us these perceptions of the body, a beautiful and terrible place, seem-

* For striking photographs of the images described here see Lennart Nilsson's *Behold Man: A Photographic Journey of Discovery Inside the Body* (Boston, 1974).

† See Marcel Bessis, *Corpuscles: Atlas of Red Blood Cell Shapes* (New York, 1974).

ingly as spacious as the sea. Within this spaciousness, though, is still solidity; the flesh has not yet resolved itself into a dew.

The moment comes now to penetrate even more deeply. To do so, we must sacrifice sight for insight. No microscope using light or even electrons can take us where we want to go. Information gained in powerful atomic-particle accelerators will be our illumination, mathematics our microscope. The power of the rational mind will provide the magnification we need on our quest.

As the magnification increases, the flesh does begin to dissolve. Muscle fiber now takes on a fully crystalline aspect. We can *see* that it is made of long, spiral molecules in orderly array. And all these molecules are swaying like wheat in the wind, connected with one another and held in place by invisible waves that pulse many trillions of times a second.

What are the molecules made of? As we move closer, we see atoms, tiny shadowy balls dancing around their fixed locations in the molecules, sometimes changing position with their partners in perfect rhythm. And now we focus on one of the atoms; its interior is lightly veiled by a cloud of electrons. We come closer, increasing the magnification. The shell dissolves and we go on inside to find . . . *nothing.*

Somewhere within that emptiness, we know, is a nucleus. We scan the space, and there it is, a tiny dot. At last, we have discovered something hard and solid, a reference point. But no—as we move closer to the nucleus, it too begins to dissolve. It too is nothing more than an oscillating field, waves of rhythm. Inside the nucleus are other organized fields: protons, neutrons, even smaller "particles." Each of these, upon our approach, also dissolves into pure rhythm.

Scientists continue to seek the basic building blocks

of the physical world. These days, they are looking for quarks, strange subatomic entities, having qualities which they describe with such words as upness, downness, charm, strangeness, truth, beauty, color, and flavor. But no matter. If we could get close enough to these wondrous quarks, they too would melt away. They too would have to give up all pretense of solidity. Even their speed and position would be unclear, leaving them only relationship and pattern of vibration.

Of what is the body made? It is made of emptiness and rhythm. At the ultimate heart of the body, at the heart of the world, there is no solidity. Once again, there is only the dance.

FOUR

What the Senses Say

Experiment, mathematics, and wild surmise have taken us to the unimaginable heart of the atom, the compact nucleus. There we have found no solid object, but rather a dynamic pattern of tightly confined energy vibrating perhaps 10^{22} times a second: a dance. This dance of infinitesimal size is incredibly energetic and heavy. If all the atomic nuclei in the body of a 150-pound man were pressed together, in fact, they would form an object no larger than a pinhead, weighing 149.9 pounds and containing the potential energy to move mountains.

The protons—the positively charged knots in the pattern of the nucleus—are not only powerful; they are very old. Along with the much lighter electrons that spin and vibrate around the outer regions of the atom, the protons constitute the most ancient entities of matter in the universe, going back to the first seconds after the birth of space and time.

What a story these dancing particles could tell us! If we could read the biography of a single proton in the skin of a fingertip, we would learn that it had lived many lives, sojourning perhaps in the petal of a violet or the viscera of an earthworm, serving to rust the sword of a Mongol warrior or enhance the blue of a Viking's eye. In prehistoric times, that same proton might have spent a few millennia making the rounds from vapor to cloud to rain to lake or sea, with side trips to such choice abodes as the tentacle of an octopus or the root of a pine tree. Perhaps it spent a few billion years wandering the lonely wastes of interstellar space, a few million heating up the nuclear blast furnace at the sun's core.

But all of this—a routine Protean life—would be of less interest to an archeologist of the nuclear world than information about the Beginning itself. For the proton is a Rosetta Stone that might help us decipher the history of the universe. The story is there, locked away not so much against the probes of our instruments as against the stubborn objective-mindedness that still denies us the experience of the dance.

There are secrets everywhere, and everywhere revelations. Turning instruments and thoughts from the infinitesimal to the cosmic, we begin to comprehend the life and times of the galaxy, basic entity of the stellar realm.

In the beginning, all the universe was thoroughly mixed, a thick soup of hydrogen gas and energy. But it is the nature of existence to eddy and swirl. Huge eddies formed in the turbulence of the primordial soup. These eddies could be stabilized by gravity if they were approximately the size of a galaxy. An eddy became an entity; embraced in the womb of its own gravity, the embryonic galaxy manifested as an enormous cloud of glowing gas. Smaller eddies within this glow began to spin, to dance, and stars condensed like whirling, burning drops of dew.

Spangled now with hundreds of billions of shining stars, the galaxies evolved, like all living things, toward increasing order and complexity. Our telescopes bring us pictures of those giant stellar organisms at every stage of growth. The youngest haven't yet been shaped up under the discipline of gravity and magnetism; they still have an irregular, unformed look about them. And there are the adolescent barred galaxies, dancers spinning gently with two outstretched arms. Then the mature dancers, spinning faster, arms beginning to curve back. And the tighter pinwheel spirals, trailing numerous glowing veils of stars as they whirl. And there are also the elderly galaxies, slow of movement: the ellipsoids and spheres.

A more explicit demonstration of the dancelike nature of the universe would be hard to imagine. It is a galactic tarantella our telescopes show us, a lively, passionate dance performed in slow motion. Interaction between galaxies is commonplace, since they are relatively close together, separated by an average distance of only about 100 of their own diameters. Like dancers set to spinning by brushing elbows with other dancers moving in opposite directions, to use Guy Murchie's lovely image, they whirl on through the further reaches of time and space, the history of their adventures written in the velocity and quality of their motion.

Galaxies sometimes meet head-on, and we might imagine hundreds of billions of stars crashing through other hundreds of billions as the penultimate catastrophe. As it turns out, however, two galaxies can pass through each other like two puffs of smoke. Since the stars within them are separated by distances of millions of their own diameters, star collisions are extremely unlikely. Thus, again, we are presented with the lovely transparency that finally defines all "objects," small and large. Whether searching for proton or pattern of energy labeled quark within proton, whether galaxy or great cluster of galaxies, we are left in the end with dancing fields of rhythm and relationship. The point of this chapter will be to show how we are connected, through our senses, to all of this.

Standing at the midpoint between proton and galaxy, we might seem to be limited to a rather meager selection of the cosmic rhythms. The full range is awesome: Our solar system takes nearly 10 billion seconds (240 million years) to make one circuit of the Milky Way galaxy. The waves at the heart of the atom whirl or vibrate 10^{22} times in one second. (That's a number with twenty-two zeros.) The whole atom, at room temperature, vibrates more slowly, around 10^{14} times a second, while molecules pulse on the order of 10^9 cycles a second.

These tempos reach far beyond the capabilities of the cells that make up our bodies. Living cells can respond to direct stimulation of up to about 1,000 cycles a second, and can manage internal cycles slow enough to match the twenty-four-hour day, the lunar month, and perhaps even the year. The complex company of cells called the brain pulses with overall wave patterns that range from more than forty cycles a second (in active concentration) down to less than one cycle a second (in deep sleep)—and that's about it.

But rhythmic waves are fluid, rather easily translated from one range to another. We are intimately connected, as we've seen earlier, with the world of sound vibrations. We are also joined, through our senses, with an amazingly wide range of rhythmic pulsations other than sound. In this regard, we can think of our senses as rhythm transformers, pure and simple.

Take the case of vision. The waves of visible light pulse between 390 trillion and 780 trillion times a second—exactly one octave, one doubling of frequency—while the cells in our body can barely make it up to 1,000. Yet we manage to bridge this gap quite nicely. In fact, the human eye can respond to a single quantum of light (the smallest amount possible) and can discern more than 10 million colors.

This feat of rhythm translation is made possible by specialized cells in the retina of the eye. The cells themselves can't respond to frequencies as fast as those of visible light, but molecules in the cells can. The responding molecules, through a complex chain of chemical reactions, send their messages to the main bodies of the cells, which then fire off the messages to the brain. To make color discriminations, as few as three types of retinal molecules are needed, since all colors can be derived from various blends of three. One kind of molecule is tuned to respond most readily to the rate of vibration we call "red," with decreasing response to frequencies on either side of that. Two other types of molecules, tuned to blue and green, also signal their graded response. The interaction of the three messages in our brain produces a sensation that we label "magenta" or "bottle green." These words, these sensations, emerge from rhythmic frequency. The fine eye, like the fine ear, possesses an exquisite sense of pitch.

Scientific investigation of how the senses work is far from complete. In registering and communicating the various rhythms that impinge upon them, the sensory systems are generally quite ingenious and sometimes maddeningly complicated. Examine the workings of the human ear, for example, and you might think there would be some simpler way to devise an organ of hearing. But the ear evolved from the gill, and its

strange mechanisms were built upon an organ designed for something entirely different. The levers
and pivots that translate the vibrations of the eardrum
into the inner ear, as a matter of fact, look like one of
those arrangements in the old Rube Goldberg car-
toons, and yet they are marvelously sensitive. The
faintest sound we can hear moves the eardrum back
and forth only 40 billionths of an inch, which is about
ten times the diameter of the smallest atom. The lever
action of three tiny bones translates this pressure to a
twenty-two-fold greater pressure on the fluid of the
inner ear.

The sense receptors for hearing are housed inside
a wedge-shaped, fluid-filled worm called the cochlea,
which coils on itself two and a half times. Tiny hairs,
about 24,000 in all, are connected to the receptor
cells. The hairs, like strings on a harp, are of different
lengths, leading early theorists to an easy conclusion:
A tone is transmitted from the eardrum to the cochlea,
causing vibrations in the cochlear fluid. Only the hairs
that are resonant to the pitch of that tone would vibrate sympathetically. The vibration of those hairs
would trigger a response in the nerve cells to which
they are attached, and these nerve cells would send the
appropriate messages to the brain.

Further research showed that the receptor cells also
respond to the pressure of standing sound waves in
the fluid of the cochlea (the best current theory), and
maybe to the bulging of the organ caused by pressure
waves, possibly even to changes in electrical potential
between different parts of the cochlea.

And this is only the beginning. Within the brain, the
auditory signal swirls around a complex circuitry with
numerous feedback loops and signal dampeners that
allow us to pick out a single conversation at a noisy
cocktail party. Our sense of hearing is indeed a won-

der, surpassing sight in many ways. When an artist blends three pigments, for example, our eyes can see the resulting blend only as a single new color. When clarinet, flute, and oboe join together, our ears can hear the resulting blend as a single new sound, and can also pick out the three ingredients within it.

It was Aristotle who came up with the idea of the five senses—hearing, vision, smell, taste, and touch—thus misleading people for some 2,500 years. Researchers at various times have identified nearly thirty additional sense-qualities associated with touch alone: pressure, contact, deep pressure, prick pain, quick pain, deep pain, warmth, cold, heat, muscular pressure, articular pressure, tendonous strain, dizziness, balance, appetite, hunger, thirst, nausea, sex, cardiac sensation, pulmonary sensation, itch, tickle, vibration, suffocation, satiety, and repletion. To simplify these "qualities" into basic touch systems, we can say the skin houses at least four separate senses: for pain, temperature, heavy and medium touch, and light touch associated with the sensitivity of hairs. Another system is embedded in the deeper tissues, muscles, and viscera.

However you classify them, the touch senses have remarkable capacities. We can easily tell the difference between a smooth pane of glass and one with invisible grooves only 1/2500 inch deep. Professional cloth-feelers can describe a fabric just by rubbing it with a stick.

It is sometimes refreshing to learn that certain ordinary human faculties have resisted every scientific attempt at explanation. And that is certainly the case with the senses of touch, temperature, and pain. There are theories aplenty but, at the basic molecular level, the precise mechanism of this sensing remains in doubt. It is one thing to say that the nerve endings associated with touch are sensitive to "the deformation of the surrounding tissue," but it is another to describe just how this "sensitivity" works. Some scientists speculate that molecules are spread apart when the flesh is pressed, thus allowing charged particles in the surrounding fluid to enter and initiate a chemical interchange that will cause the nerve cell to fire off its message. But no one is absolutely sure.

The same uncertainty exists where temperature and pain are involved. Some researchers speculate that the touch sensors respond to deformation that is horizontal to the surface of the skin, while temperature and pain sensors respond to vertical deformation. Temperature would be "measured," according to this theory, through the expansion and contraction of vertical capillaries, to which nerve endings are attached. It's interesting to note that the pathways for pain and temperature in the spinal column appear to be inseparable, and that there is a close relationship and some sort of interplay between the sites of pain and warmth, suffering and pleasure, in the brain.

The senses of smell and taste go back near the beginnings of the evolutionary journey. Creatures exist that have no hearing or sight whatever, but nowhere on the earth can you find a form of animal life that doesn't react in some manner to chemical stimuli.

The sense of smell, according to one recent theory, works on a key-and-lock principle. The odor-receptor cells, high in the back of the nasal passages, are furnished with "locks," tiny openings of just the right size and shape for certain molecules, as "keys," to fit. When the key fits properly in the lock, the cell is incited to send a message on to the brain. There are reputedly five different types of receptor cells, each type having its distinctive lock. There is one set of locks that will only receive molecules associated with a pepperminty sort of smell. There are other locks for musky, floral, camphorlike, and ethereal smell molecules. Two additional types of cells, for pungent and putrid odors, respond to the electrical charge on the molecule rather than to the shape and size. According to this theory, various combinations of the seven smell-types create all the odor sensations.

There's more to smell, however, than just the key-and-lock mechanism. Early in the nineteenth century, Michael Faraday noticed that most odorous substances tend to absorb infrared (heat) radiation. Modern researchers have updated his finding: they think your nose may act as a spectroscope, analyzing the ability of incoming molecules to absorb heat within the nasal passages and sending this information on to the brain, to be experienced as odor. A combination of the key-and-lock mechanism with the infrared spectroscope might account for the range and subtlety involved in the sense of smell.

Taste is even more of a mystery, depending not only on the chemical composition and temperature of the substance tasted, but also on its texture and smell. The basic interaction, however, is molecular, and the entire taste palette consists of four types of taste buds —for sweet, sour, salty, and bitter. To account for the gourmet's delight, you have to put these together in various combinations, and also stir in large portions of smell, vision, cultural preference, previous experience, appetite, expectation, company, and setting.

In the vibrant molecular realm, everything is rhythm and electricity. The key-and-lock metaphor is useful, but it shouldn't mislead you into thinking of a static, purely physical fit. The process is best conceived as a dancelike game with certain rules. The rules involve the interchange of electrons, ions, and atoms and, in the case of the key-and-lock situation, the correct spatial orientation of electromagnetic fields. Thus, the senses are all related. They differ simply in that they use different means for transforming fast-pulsing rhythmic waves into rates of vibration that can be handled by the brain. Certain frogs can sense light through their skins. Rattlesnakes can detect minute amounts of infrared radiation through pits between the nostril and the eye. Certain fish can detect infinitesimal electric currents through organs on the tail or near the dorsal fin. These exotic senses operate on the same principle as do our ordinary senses: response to mode and rate and pattern of rhythmic vibration.

One of the most fundamental senses in animals and humans resists categorization. That is the vestibular sense, which permits us to stand upright, to start and stop and whirl and tumble without becoming disoriented. The organs of balance consist of fluid-filled semicircular canals nestled up against the cochlea in the inner ear. The motion of fluid and calcium crystals past hair-tipped sensing cells within the canals defines the motion of the head and its relationship to the earth's field of gravity.

This delicate, sophisticated sense, according to recent research, not only helps us to stand straight, but also to think straight. The vestibular system is richly interwoven with the brain circuitry of the other senses, especially that of sight, hearing, and the various feeling senses. To understand the interconnectedness of the senses, we might consider certain oversized brain cells, "convergent neurons," that will do their job only if receiving simultaneous messages from more than one sensory system. Some of them will work only if they receive impulses from both the eyes and the vestibular system. Others need even more information, say, from the eyes, vestibular system, tactile system, and internal sensing system.

To *make sense* of what you're seeing, in other words, you sometimes need to know what the eyes register, what you're touching, your relation to gravity and motion, and the position of your joints. What we call "seeing" involves all this, and dramatically illustrates the relationship between perception and the whole body. As well, if thought involves perception, then thought is inextricably involved with the body, its balance, its ability to integrate movement and sensing and touch.

It's now becoming clear, in fact, that there is a relationship between problems of movement, balance, and touch on the one hand and problems of learning on the other. Certain children have long puzzled parents and teachers by the quirky nature of their school performance. These children often have high intelligence scores along with difficulties in specific subjects or skills. One of these bright children might have trouble telling the difference between *b, d, p, q,* and *g.* Another might comprehend well but falter in explaining. Still another might have a blind spot where math is concerned. These same children often have trouble with balance and certain types of physical movements.

A number of theorists and practitioners, notably Dr. A. Jean Ayres, are working out purely nonverbal exercises to help remedy the bothersome learning disabilities. The exercises, under the rubric of "sensory integration," aim toward integrating all the senses, with emphasis on gravity, motion, and touch. For example, children are suspended in nets hung from swivel joints. Swinging face down in the nets, they practice such skills as transferring marbles from one container to another.

One of the most effective sensory-integration procedures consists merely of brushing a child all over with a dry paintbrush. In a culture where babies are born and reared with little physical contact and where adults rarely touch except for sex, the tactile system is often starved for stimulation. And, strange as it might seem, touching, perceiving, and thinking are intimately joined.

Children with learning disabilities, and adults as well, tend to have a poor sense of rhythm. This should come as no surprise. A primary role of rhythm in any organism or society is to integrate its various parts and systems. Think of the brain as an orchestra. If any player is off-key or out of time, if any section comes in early or late, harmony becomes discord. The conscious, controlling mind (associated mostly with the cortex of the brain) can intervene like a symphony conductor and try to set things right. But this is very hard to do from above. The brain is infinitely more complex than any orchestra we could conceive. With this unimaginable complexity, the conscious mind must count on the various players and sections to harmonize and get the beat on their own. Then, too, the brain cortex has other things to do. Imagine how annoying and distracting it would be to have to figure out consciously whether the head is upright or tilted to one side. Some people are faced with just that kind of problem.

In the orchestra of the brain, the interplay of sensory input is kept in its proper rhythm, not in the cortex, but in the midbrain, which operates beneath the level of conscious thought. Messages from the various senses are blended, ordered, and cross-checked there, then sent on, if need be, to the cortex. And it is in the midbrain that we come upon one of the most remarkable features of the whole thinking apparatus: Most of the messages from *all* the senses are blended there with messages from the vestibular system before being sent on to the sites of conscious awareness in the cortex. If the vestibular system is underactive or overactive or out of tune, the effects can be pervasive and debilitating. The same problem that causes persistent (if sometimes subliminal) feelings of motion sickness can cause trouble in seeing, hearing, and feeling.

Indeed, it is the sense of gravity and balance more than any other that keeps the beat in the orchestra of the brain. East Indian musical groups generally include a stringed instrument that provides a constant droning sound, a Ground of Being against which the sounds of the other instruments can be compared. If the orchestra of the brain has a drone instrument, it is the vestibular system, which connects all our actions and thoughts with the gravity of this earth and, through that, with the gravitational field of the universe.

Just as sensing can influence thought, thought can also influence sensing. When Aristotle decided there were only five senses, he spoke with authority at a moment of high rationality in Western thought. Passed down through the centuries by the Church, his formulation has shaped human experience, and thus has prejudiced us toward what might be called *sub* natural sensing.

For it's clear that members of surviving primitive hunting and gathering bands possess sensing abilities foreign to most of us, abilities to sense underground water in desert areas, to find direction and locate position on cloudy days, and, in some cases, to "view" game animals that are out of range of the conventional senses. Verification of these abilities under controlled laboratory conditions is close to impossible, in that it would involve the trauma of taking primitive people out of their home environment. The feats, however, have been observed in the field by respected witnesses.*

Must such abilities die with the last remnants of the primitive bands and tribes, or might they be regained by members of present cultures? In science, as in everything else, the answers you get depend to a large extent on the questions you ask and the way you ask them. When our prejudice against the extrasensory is put aside, questions can be asked in such a way as to yield some interesting results.

For example, a Sorbonne physics professor, Yves Rocard, took the idea of water dousing seriously enough to devise experiments in dousing for "average" subjects. He had these people hold the muscles of the arm taut while balancing a long stick and searching for underground water. According to Rocard, changes in the local magnetic field brought about by the water in the soil caused the muscles to relax and the stick to dip. Professor Rocard also planted electrical coils underground to approximate the magnetic changes caused by water. By using a magnetometer, he discovered that his subjects could detect changes in the magnetic field as small as .3 milligauss.† By comparison, the earth's magnetic field at surface level in

* See Robert L. Van de Castle, "Anthropology and Psychic Research," in *Psychic Exploration,* Edgar D. Mitchell, ed. (New York, 1974).

† Yves Rocard, "Actions of a Very Weak Magnetic Gradient: The Reflex of the Dowser," in M. F. Barnothy, ed., *Biological Effects of Magnetic Fields,* I (New York, 1964), pp. 279–86.

the temperate zones is 500 milligauss, which might make you think that "average" people could also devise some way to detect magnetic north. Indeed, this seems to be the case. Some participants at my workshop sessions learn to find magnetic north blindfolded by scanning with their hands.

The human body, like the bodies of all living things, creates its own electromagnetic field. Every cell contributes to this field, especially the active gland cells and the muscle cells, which produce relatively strong electrical current upon each contraction. The nervous system is a network of unceasing electrical activity; the brain, an incredibly complex switchboard on which every light is twinkling, night and day. The electrical activity within the brain, as we have seen, is organized into pulsing waves which can be measured on the surface of the scalp, and which also can propagate out into space at the speed of light. Similarly, the heart and its extended system of blood vessels produce electrical current and magnetic force with an accompanying electromagnetic field. The current generated by the heart itself can be measured on the surface of the chest as a charge of as much as a hundredth of a volt. The electromagnetic field associated with the cardiovascular system has been detected by sensitive instruments several feet away from the body.

Each of us, then, is a radio transmitter. The signal strength is minute compared with that of even the smallest walkie-talkie, but it is not inconsiderable. With the proper tuning and amplification, the human signal might be received and decoded at a distance. The question remains, can we also operate as radio receivers? The conventional answer is a definite no. Standard texts on the senses state flatly that we cannot feel electricity as such, nor do we have any sense or-

gans for electromagnetic radiation other than that in
the bands of radiant heat and visible light.

The standard texts, however, reflect our present
limited knowledge on the whole subject of sensing.
The senses, as it turns out, rarely if ever operate as
separate and discrete systems. To be meaningful, sensory information is modulated and blended and
strongly influenced by prior experience, preconceptions, and cultural taboos. On some occasions, the
senses take each others' places. Sounds of very low
pitch (less than sixty cycles a second) are "heard" not
so much through the ears as through the skin. An
electrostatic field such as is produced by pulling two
sheets of plastic apart can be sensed through touch,
since it causes the hairs of the body to move. The
mysterious sense of "presence" that sometimes an-
nounces the approach of another person in the dark
might operate through this mechanism, since the
human body produces an electrostatic as well as electromagnetic field. But there's probably more to it than
that. In aikido classes, I've frequently witnessed feats
of sensing the presence of another person at a distance that can't be explained in terms of electrostatic
or heat sensitivity. We should probably assume that
our scientific understanding of sensing is still in the
rudimentary stage.

We should also bear in mind that electromagnetic
radiation can bypass the sense organs and affect the
cells of the body directly. Dr. W. Ross Adey, head of
UCLA's Brain Research Institute, has shown that
a weak, pulsating electric charge applied to metal
plates several inches on either side of a person's
head can significantly alter the ability to estimate
time, especially if the pulsations are in the range of
seven cycles a second. Experiments of this type

on cats and monkeys have revealed changes in brain chemistry and electrical activity as well as time sense.*

This is a matter of some concern, since our usual metropolitan environment bombards us with electromagnetic radiation from radio and television transmitters, microwave relays, radar, high-tension lines, home appliances, and, perhaps most insidious of all, from the network of ordinary sixty-cycle house wiring within which we live. When enough radiation of certain frequencies passes through our bodies (more than normally caused by the above), the molecules that make up the cells are vibrated so violently that they start heating up. This kind of radiation, as a matter of fact, is used to cook meat in microwave ovens.

We are affected many ways, known and unknown, by the pulsing fields that crisscross all the space in which we live. Our health and well-being are influenced to some extent by our connection to the rhythms of the tides, the seasons, and the turning of the earth. There is reason to suspect that we are touched by even more distant events. The cycle of sunspots (giant solar flares that send out showers of cosmic rays) affects radio reception and seems to influence weather and crop yields. It has also been suggested that it corresponds with aspects of human behavior: for example, with outbursts of excitement among patients in mental institutions, and indeed with the mass human excitability that leads

* R. J. Havalas, D. O. Walter, D. Harver, and W. Ross Adey, "Effects of Low-Level, Low-Frequency Electric Fields on EEG and Behavior in *Macaca Nemestrina*," *Brain Research* 18 (1970), pp. 491–501. Also see recent unpublished studies by Adey, c/o Brain Research Institute, University of California, Los Angeles, CA 90024.

to riots and revolutions.* As will become clear in later chapters, we are indeed *in relationship* with all that is.

At the height of the nineteenth century, Western man envisaged himself in a heroic stance, an isolated individual eternally at war with Nature and the "heathen hordes." The body was seen as a fortress, the five senses as sentinels to warn against dangers in the inevitably hostile outside environment. That vision shaped perception and reality. The outside world was hostile. There was always war.

That age has passed. How arrogant it was to think that we could stand alone and apart, against the world. Now we are beginning to see that the senses—all of them in all their interplay—are not mere sentinels but means of connection. By transforming rhythms to a realizable range, they give us information about relationships, so that we can relate more harmoniously with our world. What the senses say is that we are not apart from, but a part of all that we perceive.

* See Edward R. Dewey, *Cycles: The Mysterious Forces That Trigger Events* (New York, 1971), pp. 57–59; also Gay Gaer Luce, *Body Time* (New York, 1971), p. 51.

FIVE
Personal Identity and the Inner Pulse

The chance of two fingerprints being exactly the same is said to be less than one in 64 billion; matching a whole set of fingerprints with another whole set is impossible. Faces are equally distinctive and, except in the case of identical twins, easily recognizable. A good voiceprint—a printed replica of electronically recorded voice frequencies—effectively identifies the speaker. Handwriting can be forged, but only through the most consummate skill, and then not with unshakable fidelity. A bloodhound can pick up the scent of one person out of a million. Brain-wave patterns, as our means for analyzing them become more precise, are seen to be entirely distinctive. Human infants are born with unique and identifiable rhythms of sleeping and waking; recent research suggests that a newborn's breathing pattern is as distinctive as a thumbprint.*

* See Evelyn B. Thoman, "Early Development of Sleeping Behaviors in Infants," in *Aberrant Development in Infancy: Human and Animal Studies*, N. R. Ellis, ed. (New York, 1975), pp. 123–38.

Personal identity is, in fact, an essential quality of human existence, and its operations as expressed in a rhythmic inner pulse will be seen as central to the thesis of this book.

The ability to recognize other individuals as members of the same species goes far back in evolution, certainly as far as the development of separate sexes. Social insects such as bees, ants, and termites have the additional ability to distinguish between different castes (queen bees from workers, for example), and also can recognize one life stage from another (eggs from larvae, and so on). Going up the evolutionary ladder, all vertebrates can distinguish among infants, juveniles, and adults of their own species; but true individual recognition arises, with a few exceptions, only at the level of birds and mammals.

The brain/sense mechanism, characteristically, comes up with varied and ingenious means for the recognition process. Some primates can recognize faces, just as people do. Many mammals, as every dog owner knows, use secretions as personal signatures; the ability of these animals to distinguish one individual from another by smell is phenomenal. The howl of a wolf communicates its emotional state, its location, and its separate identity; experiments with recordings and sound spectrographs show that wolves can make out extremely subtle differences in sound.* Birds can distinguish individuals through the absolute frequency or pattern or "dialect" of their songs or calls, as well as through visual appearance.

Individual recognition plays a role in staking out territories; a songbird knows and accepts the cooperative competitors on the borders of its territory, but

* John B. Theberge, "Wolf Music," *Natural History* LXXX: 4 (April 1971), pp. 37–42.

becomes upset at the arrival of a stranger. Recognition is also a key factor in pair bonding, in the rearing of young, and in establishing the pecking order that provides the social glue in many social groups.

All human societies, of course, are based around personal identity, and it is instructive to note the extreme concern with which this matter is treated from birth to death. There is something nightmarish and almost unthinkable about a human being without an identity. Since every cell in the body (except the blood cells) contains molecules of DNA on which is written the blueprint for the entire body, it is theoretically possible to create an exact copy of any human being —or a million copies—from one cell. Cloning, as such a process is known, is not only a staple of science fiction, but is actually under consideration by some scientists. Still, our common sense rebels at the notion, and we may find ourselves wishing it will turn out to be more difficult than some futurists believe, perhaps impossible. For we all realize at the deepest level of our intuition that a million John Smiths, even a million Albert Einsteins, would be something other than human. Recent research, in fact, suggests that the evolutionary process itself involves mechanisms for producing and maintaining individual variations within each species.*

To be human, it seems clear, is to have a personal identity. This identity is unique and irreversible. It provides our particular viewpoint on the universe. It expresses itself in numerous ways, subsuming what we call body, mind and spirit, memory and works. It is the creative artist's finest offering. It survives death.

---

* Bryan Clarke, "The Causes of Biological Diversity," *Scientific American*, (August 1975), pp. 50–60.

The little town of Monroe, where I spent my four-teenth summer, seemed miles from everywhere, a mythic place on the gently rolling red-clay earth of northern Georgia. There was one movie theater, no air conditioning, no television—only voices, the songs of birds, the hum of insects, and the pervasive heat. And it was there one morning that my older cousin gave me *Look Homeward, Angel* by Thomas Wolfe and insisted that I begin reading immediately.

Four hours later, at the height of the afternoon heat, I let go the book, hands trembling, face flushed. I had finished only some fifty pages, and my life had been changed. I was shaken, not so much by the spe-cific content of the writing as by the quality—the rhythm if you will—of the experience. You may want to call it a shock of recognition; it is true I met myself on those pages. But I also met Thomas Wolfe. *In person.*

As I went on through the book, forcing myself to read slowly so that it wouldn't end, he was there with me, not a mere literary presence, but somehow real: Wolfe himself. Years later, on a trip to Asheville, North Carolina, I visited Wolfe's home and grave. I met people who had known him intimately, and had lunch and dinner with his sister. But his personal pres-ence was not so well rounded and clearly defined on that trip as it had been during those long, hot days and magical nights in 1937, when I first read *Look Home-ward, Angel.* I knew Wolfe then through an aspect of his being that was somehow encoded in the pages of his book.

The record drops to the turntable and begins to turn. The tone arm lifts, moves sideways with disinter-ested mechanical grace, hovers for a moment over the edge of the record, then descends. The speakers crackle loudly and I realize the record is more badly scratched than I'd remembered. A single muted violin begins the quartet. The tone seems distorted. Maybe something's wrong with the needle. The second violin joins in, the viola and cello.

And once again the miracle happens. The flaws are forgotten. Ludwig van Beethoven is here in the room. He is addressing himself to me alone. I *know* him. Across 150 years, transmitted by lines and dots on paper, put into sound by four men playing wooden instruments in Boston, translated into electronic impulses and thence into the mechanical vibrations of a stylus on a master disk, mass-produced into records, translated back into electricity then sound, Beethoven's personal presence survives. Ten times removed in time and space and intermediaries, his *C-Sharp Minor Quartet,* Opus 131, still contains him, *is* him.

Something of Wolfe, his "actual" presence, was contained, not so much in the content of his writing as in the cadence, the spaces between the words, if you will. Beethoven's presence was encoded and realized through an entirely different process, again a rhythmic one. My most striking experience of both these men, no longer living, was that of their personal identity, of whatever it was that made each of them unique in all the universe.

This feeling that the artist is contained in the work is a familiar one. But we have been taught since earliest childhood to distrust our feelings. What we need is something objective—a reading on a dial, a line on a graph—to let us trust what we already know. In the case of Beethoven and other musical composers, there is now, fortunately, an objective measure of personal identity and presence, discovered by a musician-scientist named Manfred Clynes.

First of all, Clynes has developed a precise way of defining and measuring certain emotional states. These states, he finds, are extremely specific. According to Clynes, "the qualities of the spectrum of emotions are more precise by far than the words used to describe them." Each of seven basic feeling states—anger, hate, grief, love, sex, joy, and reverence—has its own characteristic gesture, which he calls its "essentic" form. This gesture, preprogrammed in the brain and expressed through the muscles, can manifest in any number of ways: through a facial or vocal expression, a dance movement, the stroke of a pen, or even through the pressure of a finger. It is by means of finger pressure, in fact, that Clynes is able to record the exact form of each emotion or "sentic state."

His method is simple. He asks his subject to sit in a comfortable position with the right middle finger (if the person is right-handed) resting on a button that is hooked up to measure both vertical and horizontal pressure. The subject is asked to express, say, anger with a single application of finger pressure. This is repeated thirty to fifty times, and the results are averaged by a computer in order to cancel out minor fluctuations. The product is presented graphically by the computer as a pulse, a form against time.

The essentic form for anger looks like this, with the top line representing the vertical finger pressure, the bottom line the horizontal:

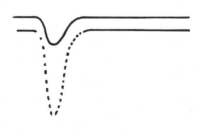

The pulse of love is quite different, as follows:

The graphic representation for reverence *looks* like reverence:*

Essentic form seems to vary little from culture to culture or between the sexes. Members of various cultures Clynes has tested—American, Mexican, Japanese, Balinese—show an amazing consistency in their responses, as do men and women. Nor can people be trained or conditioned to switch essentic form. A jabbing motion can't replace a caress in expressing love, and vice versa. Any attempt to teach such perverse forms of expression seems simply to block the ability to feel. An emotion that seems to be experienced in the "mind" is realized in gestures, movements of muscles, breath, voice. Suppression of gesture serves to repress emotion. When we are taught to block the expression of grief, for instance, it is the force of life itself that is stifled.

Imagining anger or love or joy makes it possible to express the essentic form for the emotion in question. The process also works the other way. Expressing essentic form through gesture—even by doing some-

* Drawings from Dr. Manfred Clynes, *Sentics: The Touch of the Emotions* (Garden City, N.J., 1977), pp. 35–39, 93–94.

thing as simple as pressing a button—increases the intensity of the emotion and creates significant changes in heart rate and respiration. Realizing this, Clynes has developed a practice that involves going through the seven basic emotional states in order, using fantasy and finger pressure. The purpose could be called therapeutic or, better, transformational. Following instructions on a cassette tape, you can experience a "sentic cycle" in thirty minutes, with possible benefits that range from cathartic release to increased creativity.* I have employed these cycles at sessions at my aikido school with reliably good results.

It is in music, however, that essentic form reaches its highest expression. And it is here that Clynes has made what for me is his most exciting suggestion. Music is an ideal medium for the expression of feelings. As Mendelssohn pointed out, music cannot be expressed in words, not because it is vague, but because it is *more precise* than words. A great musical performance is actually an exercise in the accurate and exquisite communication of emotions. Music is not rote. For the audience to feel the emotion in a musical phrase, the performer must also feel it, in mind, body, and spirit. Exaggeration, seductiveness, falsification, become immediately apparent to any discerning listener.

There is something else about great music: it seems to contain, as I've suggested, the composer's voice; it evokes his personal presence. Clynes reminds us that one can't take a phrase from Mozart and put it into Beethoven—even if it's the same, note for note—without stumbling over the beat. There seem to be two

* Sentic-cycle tapes or records may be obtained from Secretary, American Sentic Association, Box 65, Palisades, N.Y. 10964.

distinctive and quite different underlying musical pulses involved.

Seeking a measure of the underlying pulse, Clynes invited well-known conductors to "conduct" pieces of music by famous composers. They did this simply by thinking each piece while repeatedly pressing the button of Clynes's sentograph. The conductors would think of pieces by Beethoven and press the button fifty or so times for each piece. The computer would average the responses and print the resulting pulse. Whether the conductor was Casals, Serkin, Perahia, or Clynes, whether the piece was the second movement of Opus 13 or the third movement of Opus 109, Beethoven's musical signature looked like this:

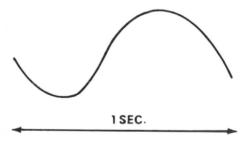

1 SEC.

The pulse for Mozart—whatever the piece, whoever the conductor—turned out to be quite different and equally distinctive:

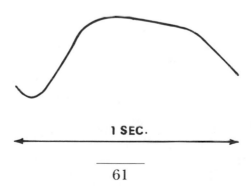

1 SEC.

Clynes went on to seek the inner pulse, as he called it, in the music of other composers. He found it to be remarkably stable over time and space. Like the color red, "the personality Beethoven revealed in his music has a precise existence that continues. This stability is not a result of 'tradition' or of 'style.' Mozart and Haydn have very different inner musical pulse shapes! So have Debussy and Ravel."*

Like a fingerprint, a voiceprint, a signature, or a chromosome, the inner pulse is an expression of identity. Clynes's investigations of the inner pulse have been confined to the field of music. But if the inner pulse is like other essentic forms, it must have innumerable modes of expression, not music alone. Thus, we can say that Thomas Wolfe expresses his inner pulse through the rhythm and the power of his pen, and that his living presence can be sensed in a dog-eared book just as Beethoven's can in a worn-out record.

Clynes's work is not definitive, but it suggests that each of us has our own inner pulse, an essential aspect of our being that finds expression whenever we walk, speak, sing, write, shake someone's hand, play golf, make love—a quality not yet measured but available for eventual identification and measurement. Perhaps this is what mysteriously invites or repels, what identifies us on a deeper level than the contours of the face. In this light, a gentle touch can announce the inner workings of the mind and spirit, and it is more than mere irrationality that leads us to fall in love in a flash while dancing with someone.

Each part, each aspect, in some way expresses the whole. Thus, the inner pulse has a special significance.

* Clynes, *Sentics*, p. 94.

For it might well provide us one neat summation of all the pulsations, the weaving play of wave upon wave, that constitute a human individual.

The question of personal identity has tended to lead philosophers directly into the trap of mind-body dualism. "Should the soul of a prince," John Locke wrote, "carrying with it the consciousness of the prince's past life, enter and inform the body of a cobbler, as soon as deserted by his own soul, everyone sees he would be the same *person* with the prince, accountable for the prince's actions." As in this puzzle story, the problem of reidentification has always been posed as one of mind or memory versus body, with mind generally seen as the primary criterion. But when we look at a human individual as a series of rhythmic patterns or wave functions, summed up as a single wave function or inner pulse,* then the dualism of mind and body dissolves. The body can be seen as one manifestation of the inner pulse, with its various rhythms, even to the frozen rhythms of fingerprints, growing out of the distinctive wave function. Mind as memory is seen as another manifestation. Thus, separating mind and body is theoretically as well as practically impossible.

Indeed, in terms of this speculation, it is the inner pulse that is stable and persistent at the most fundamental level: the unitary identity that explodes out into the world as a multiplicity of identifying characteristics. Thus, if the physical body is one manifestation of the inner pulse, so are the fields in and around the physical body, whether the scientifically verifiable electromagnetic/electrostatic fields or the putative

---

* This inner pulse, in toto, is of course far richer than the sample that represents its expression in the music of the classical composers.

mystical fields—the "radiant body," the "astral body," and so on. If mind as memory is one manifestation of the inner pulse, so is mind as intellect. And the span of a whole life, if only we had ways beyond intuition of charting it, would be seen to have a distinctive rhythmic pattern, a pattern that would by no means be erased upon the death of the physical body, but would persist in the everlasting web of rhythmic relationships that constitute a meaningful universe. And in this life, if such a thing as an out-of-body experience exists, the disembodied entity would surely be a function of the same essential pulse as that of the physical body. And spirit or soul, whatever endures beyond works and memory, would also be subsumed in the quality of identity that expresses itself in distinctive rhythm.

We can look at personal identity, then, as unique, irreducible, and irreversible—one of the essential qualities of being. There is another quality, one that binds identity in a paradoxical relationship containing, in potentia, almost infinite power. This quality—I shall call it "holonomy"—is even stranger and more mysterious than identity, as we shall see in the next chapter.

SIX

"Holonomy": The Web of Existence

No one way of viewing the world guarantees the solution to its mysteries. Each vantage point offers a truth of its own, bringing to light new beauty and utility, creating new puzzles and contradictions. Newtonian science provided a reliable framework on which the modern West could be built. But then the twentieth century took another look and discovered that the Newtonian framework is only a special case in a larger reality. Space, according to relativity theory, is not fixed, nor is time steady. Common-sense notions about cause and effect don't hold in quantum theory, and it now appears that objects are to be known by their relationships rather than by any independent fixed character.

Thus far in this book, I've presented a variation on that theme, choosing to view the world mostly in terms of rhythmic fields, with music and dance as perhaps its most appropriate metaphors. I've introduced the idea of an individual human identity that can be expressed through a distinctive inner pulse. I've suggested that the body—bones, flesh, and all—is ultimately emptiness and rhythm. And I've treated the senses as means of connecting the organized rhythmic fields that we call the self with all the rhythms of the world, and transforming this material into the picture or projection that we call ordinary reality. What I want to do now is explore the possibility that our connectedness with the world goes beyond the sensory; that it is built into our very structure. This exploration begins with a hard-bitten, wiry Irishman named Pat Price.

I met Pat Price in October of 1973, at the height of the Israeli-Egyptian Yom Kippur War. We were both visitors at the Stanford Research Institute (SRI), a large private organization near Stanford University devoted to research on a wide variety of subjects for governmental and industrial clients. I was there primarily to check on an experiment in remote viewing. Price was there as one of the subjects in the experiment. Circumstances threw us together for most of a morning and for lunch, during which time we discussed the remarkable abilities he claimed for himself.

It should be understood that Pat Price was no hazy, mystical type. He had been a police commissioner and city councilman in Burbank, California, and his words came out in the staccato, no-nonsense rhythm of an old Irish pol. Early in our talk we discovered our common background as World War II combat pilots. He said he had flown low-level missions in B-25s. I told him I had instructed in B-25s, then had gone to the Southwest Pacific to fly low-level strikes in A-20s. Price thought I'd be interested in what he had been doing over the past few days.

"I've been flying missions with Israeli pilots," he told me. "Sitting right behind them in the cockpit. It's a helluva war."

He described several missions in detail—a strike

against oil storage tanks, a number of battles against Egyptian tanks—all of which he had "flown" while remaining in the vicinity of SRI. "Hell, it's not much different from the stuff we did in B-25s. The ground's coming up faster. You're firing rockets instead of 50-calibers. But it's the same stuff."

I asked him about the flight characteristics of the Phantom jet, and found myself engaged in a matter-of-fact technical discussion about instrumentation, control pressures, and ground fire.

"There's that turbulence when you go through the smoke of the explosions," he said. "One time, just as we had a tank lined up in our sights, there was a big cloud of smoke and the pilot couldn't see the tank, but I could. So I gave him a little nudge and he fired the rocket. We got the tank."

Price explained that he could go wherever he wished in time or space, that he could "see" distant events with great clarity. "Everybody could do it," he assured me, "but, you see, the first rule we agree to before we're born is that we forget all this knowledge. I'm one of those guys who doesn't play by the rules."

I was compelled by Price's confidence and impressed by his command of detail. But I remained skeptical, especially when he spoke of "helping" an Israeli pilot bag a tank. And indeed it turned out that, under the rigidly controlled circumstances necessary for scientific proof, his distant vision was not as flawless as he claimed. In one experiment, for example, he described a distant target area as including two water filtration pools, when actually the bodies of water in question were swimming pools. Nevertheless, he scored seven "direct hits" for the nine target areas selected in the experiment.

Price was one of six persons tested in the first series of remote-viewing experiments reported out of SRI.

The chief experimenters, Harold E. Puthoff and Russell Targ, knew that, as with all experiments in the paranormal, their controls and procedures would be picked to pieces by critics. That's exactly what has happened since the publication of their results in the March 1976 *Proceedings of the Institute of Electrical and Electronic Engineers.* But every objection has been refuted to my satisfaction. What's more, the SRI work has been independently replicated several times by other research organizations, and now must stand as one of the most serious challenges yet to the conventional Western view of human capabilities.*

In brief, the experiment works this way: Pat Price (to take him as an example) is closeted with Russell Targ in an electronically shielded room at SRI to wait thirty minutes before beginning to describe the remote location. After they are closeted, Harold Puthoff goes to the office of a senior SRI official and is given a sealed envelope containing travel instructions to a target location. Previously, the official has prepared a pool of 100 such locations and has made himself responsible for keeping the locations secure. The sealed envelopes are arranged in random order. Puthoff is given one of the envelopes. He takes it to his car, then opens it. Accompanied by another SRI employee, he drives to the target site. The two of them remain there for an agreed-upon fifteen minutes following the thirty minutes allotted for travel.

Targ, the experimenter remaining with Pat Price, has no knowledge of the particular target or of the pool of targets. During the fifteen-minute period when Puthoff is at the target area, Price describes to

* See Harold E. Puthoff and Russell Targ, "A Perceptual Channel for Information Transfer over Kilometer Distances: Historical Perspective and Recent Research," *Proceedings of the IEEE* 64:3 (March 1976), pp. 329–54.

Targ what he is "seeing" at Puthoff's location, and makes drawings if he wishes. The conversation is tape-recorded and transcribed. Later, Price is taken to the site for discussion and feedback.

After all nine of Price's remote-viewing tests are completed, the transcripts and drawings are given to an SRI research analyst not otherwise associated with the project. The analyst goes to the nine target sites and ranks the transcripts and drawings in order of what he considers the best matches, on a scale of one to nine. Since neither the experimental subject nor the analyst has any knowledge of the correct answers, the experimental design is what is known as "double blind."

In Price's case, as we've seen, the analyst made perfect hits on seven of the nine targets, giving the other two a third and sixth rank. The odds of such a thing happening by chance are one out of 29,000. To double-check these results, a panel of five additional SRI scientists not associated with the project were asked to blind-match the transcripts and drawings with the locations. This judging procedure also produced highly significant results.

Puthoff and Targ went on to test six more subjects in the same way. Three of them, including Pat Price, were experienced in remote viewing, while three were classified as "learners," people who had never before attempted the skill. One of the learners, a professional photographer named Hella Hammid, came up with even better results than Price's. The analyst correctly matched five of her descriptions and gave a second ranking to the other four, for odds of one out of 180,000. The other two experienced subjects also produced excellent results, while the two remaining learners did not score significantly better than chance. But one of them, four minutes into her very first attempt, recognized the target location and identified it by name. An excerpt from the unedited transcript gives a feeling of the process. Phyllis Cole is the subject and Russell Targ is the experimenter with her. It should be restated here that Targ has no knowledge whatever of the target.

TARG: Can you move in where he [Puthoff] is standing and try to see what he is looking at?

COLE: I picked up he was touching something—something rough. Maybe warm and rough. Something possibly like cement.

TARG: It is twenty-four minutes after eleven. [The remote viewing conversation began at 11:20.] Can you change your point of view and move above the scene so you can get a bigger picture of what's there?

COLE: I still see some trees and some sort of pavement or something like that. Might be a courtyard. The thing that came to mind was it might be one of the plazas at Stanford campus or something like that, cement . . . Some kinds of landscaping . . . I said Stanford campus when I started to see some things in White Plaza, but I think that is misleading. . . . I have the sense that he's not moving around too much. That it's in a small area . . . I guess I'll go ahead and say it, but I'm afraid I'm just putting in my impressions from Stanford campus. I had the impression of a fountain. There are two in the plaza, and it seemed that Hal was possibly near the, what they call the Mem Claw.

TARG: What is that?

COLE: It's a fountain that looks rather like a claw. It's a black sculpture. And it has benches around it made of cement.

As it turned out, everything Phyllis Cole said was exactly right.

From these and other experiments, Puthoff and Targ have come up with certain observations. Generally, the subjects do better when describing shape, form, color, and material rather than when trying to analyze or name what they are seeing. Their drawings tend to give better clues than do their words. Curiously, objects in motion at the remote locations—a train crossing a trestle, for example—are rarely mentioned. And yet, the subjects are able to shift their points of view so as to "see" things that are invisible to observers on the spot. Distance seems to have no significant effect on the results, nor does the presence or absence of electronic shielding.

It's as if the subjects can pick up elements of the scene anywhere within view of the on-site observer, and then some. Details, however, are often fuzzy or entirely missing. What comes to mind here is a photograph in which the images are poorly resolved—and this idea will play a part in our speculations as to how remote viewing works.

First, however, we must deal with the project's most mind-boggling discovery. During the course of the experiments, subjects occasionally told the experimenter remaining with them at SRI that they had had images of the target location even before the experiment began. In two cases, these precognitive descriptions were exceptionally accurate. This led Puthoff and Targ to design an experiment that would involve viewing that is remote in time as well as in space.

The experiment was worked out so that the subject, Hella Hammid in this case, would finish her description five minutes before the target was selected and twenty minutes before the observer arrived at the target location. Stringent controls assured there would be no leakage of information. Four of such tests were carried out, with results that were the best yet. Hella

Hammid's descriptions of the locations that Hal Put-hoff was *going to* visit were coherent and compelling. Three independent judges matched all four of her descriptions to the correct target locations.

In their report, Puthoff and Targ confessed a reluctance to publish these results "because of their striking incompatibility with existing concepts." They did so for what they call "the ethical consideration that theorists endeavoring to develop models for paranormal functioning should be apprised of all the observable data." They also pointed out that nothing in the fundamental laws of physics forbids the transmission of information from the future to the present. Mathematically, the familiar field equation for electromagnetic fields can be solved, not just for a retarded time, but also for an advanced time, implying that a field (such as a radio signal) can theoretically be observed *before* it has been generated by its source. This, of course, would turn the familiar chain of cause and effect around, which offends our common sense. But stranger things have been predicted and then observed in the fantastical world of modern physics.

Indeed, it is to physical theorists that Puthoff and Targ (themselves physicists) turn for possible explanations for the phenomena revealed in their experiments. One hypothesis, proposed originally by a Soviet scientist, holds that extremely low-frequency waves carry the information necessary for remote viewing. There are, in fact, Schumann waves that resonate at 7.8 cycles a second in the channel all around the globe between the earth's surface and its ionosphere, several hundred miles up. This falls within the frequency range of the brain's theta waves—and the rate of musical vibrato. Thus, it might be that Schumann waves are sometimes "hooked into" the pulsing of human brains, connecting them at a distance. The connection would be one of resonance rather than radiation, which means that signal strength would not fall off in the usual way over the miles. Nor would these waves of extremely low frequency be blocked by electronic shielding. The problem is that waves vibrating so slowly can carry very little information in a given time, perhaps not enough to account for the detailed descriptions provided by some remote-viewing subjects.

More compelling explanations emerge out of recent interpretations of quantum theory. Many physicists are now coming around to the point of view that there must always be significant connections between the observer and the observed, between consciousness and the environment. It's quite clear that observation affects the microworld of electrons, protons, and so on. To observe precisely the position or momentum of an electron is to change its position or momentum. Something like this also appears to be true of the larger world. In the words of the distinguished Princeton physicist John Wheeler, "I think that through our own act of consciously choosing and posing questions about the universe we bring about in some measure what we see taking place before us."*

Not only does quantum theory suggest a connection between consciousness and the environment, but also between every particle in the universe. A 1964 formulation by physicist J. S. Bell, known now as Bell's Theorem, emphasizes that "no theory of reality compatible with quantum theory can require spatially separated events to be independent," but rather must allow for interconnectedness of distant events in a way that's contrary to ordinary experience.

* See "The Princeton Galaxy," interviews by Florence Helitzer, *Intellectual Digest* (June 1973), p. 32.

All this is to say, hard as it might be to understand, that mind and matter can and do influence each other, that everything that happens everywhere is somehow connected, that perhaps God's awareness of the sparrow's fall is, in the ultimate sense, just the way things are.

The question remains: Even if everything is connected, through what sense or sensory system can we experience the connection? Or can we "sense" things and events, not through our senses at all, but rather through our own internal structures? To examine this possibility, I want to take a few pages to introduce a process called holography.

In every lifetime, there are moments of sudden illumination. Some are earthshaking, involving supreme effort or a brush with death. Others are less strenuous: a new idea, the discovery of the single piece that makes a puzzle fit together. Learning about the hologram was such a moment for me. The idea appeared in an article in the June 1965 issue of *Scientific American* about a radically new photographic process.* In this process, the object to be photographed is illuminated by light from a laser. Unlike ordinary light, laser light is coherent; that is, it marches along in orderly, matching waves. When these waves strike the object, they reflect back off every point of the object as spherical waves that reflect the shape of the object. The spherical waves are like the concentric waves you see in a pool of water, except three-dimensional instead of two-dimensional. A piece of photographic film is exposed to these waves; no lens or camera is needed. The film is also exposed to a reference beam—light

* Emmett N. Leith and Juris Upatnicks, "Photography by Laser," *Scientific American* (June 1965), pp. 24–35.

from the laser that is reflected back from a mirror rather than from the object being photographed. The interaction of the two sets of light waves on the sheet of film produces an interference pattern similar to the varying patterns of dark and light you see when you hold two mesh screens together at different angles.

When the film is developed, it bears no resemblance to the object that was photographed. In fact, it looks like "a hodgepodge of specks, blobs and whorls," a collection of oversized fingerprints. What's happened is that all the wave-fronts of light coming off the object have been captured on film to produce what is called a *hologram*.

How can you get a picture from the hodgepodge on the film? Simply by shining a beam of laser light through it. The captured waves are released and the image of the original object is reconstructed, seeming to hang in midair in perfect 3-D. Actually, the image is even better than ordinary 3-D. When you move your head from side to side while looking at it, you can see *around* things, bringing to view parts of the scene that might otherwise be hidden behind. The hologram, that confusing pattern of swirls, contains considerably more information than does a conventional photograph.

What's more—and this is the bit of information that provided me a moment of delight—when a hologram is cut in half, the entire picture is contained in both halves. When these halves are cut in half, the same thing is true, and so on down to the smallest fragment. Each part of the hologram, no matter how small, can reproduce the whole image when illuminated by laser light. As the pieces get smaller, the reconstructed image gets fuzzier. *Resolution* is lost. But the whole picture is there. This marvelous property is explained by the fact that each point on the hologram receives light from all parts of the subject and thus contains, in an encoded form, the entire image.

That the whole of a thing is contained in each of its parts is by no means a new idea. As we've seen, the essence of a whole person is expressed in certain of his or her parts—fingerprint, voiceprint, pattern of respiration, inner rhythmic pulse. In addition, the DNA in the nucleus of every cell contains the blueprint for the whole body. But the idea of all-in-oneness that can be drawn from the hologram goes beyond that. It is closer to the Hindu idea of the Net of Jewels, in which every jewel, every piece of the universe, contains every other piece. It brings to mind the oft-repeated Hindu phrase, *tat tvam asi* ("Thou art That"), which signifies that God or the Ground of Being is immanent in all of creation, including the individual self, and that this oneness can be directly experienced through spiritual discipline. Indeed, the mystic tradition of every great religious tradition, be it Hindu, Buddhist, Islamic, Jewish, or Christian, insists that every part of the universe in some sense contains the whole. It was a Christian mystic, William Blake, who wrote,

> *To see a World in a Grain of Sand,*
> *And a Heaven in a Wild Flower,*
> *Hold Infinity in the palm of your hand,*
> *And Eternity in an hour.*

Our common sense bridles at the idea of each of the diverse fragments of the natural world—grain of sand, seed, leaf, stone, person—actually *containing* the entire universe. Yet this is perhaps the most compelling knowledge gained by those who achieve the highest states of contemplation. In the hologram, we see science once again imitating mysticism, presenting us with a modern device which helps explain an ancient truth.

Since 1965, in fact, holography has developed and expanded in many directions, suggesting that the method can be used for encoding and storing all kinds of information. Sound waves have been used holographically to create visual images of the internal organs of the human body. Holography has been applied to microscopic and X-ray photography. Experimental movie and television systems have been developed, and there are now holograms made with ordinary light rather than with lasers. As it turns out, even the reference beam that produces the interference pattern isn't needed. Whenever two or more sets of waves intersect, holography is possible. Since every particle in the universe warmer than absolute zero is constantly producing fields of waves, and every organized combination of particles is also broadcasting its own unique fields, the number of intersecting sets of waves approaches infinity. Theoretically, some sort of superholograph could be made at any spot in the universe containing information about the whole universe from that vantage point. But before taking it quite that far, let's look for a moment at the convoluted mass of tissue inside our own heads.

The human brain is the most complex entity we have yet discovered anywhere, and it now appears the brain itself, whether human or animal, may operate in a holographic fashion. One of the boldest of brain theorists, Dr. Karl Pribram of Stanford University, has come up with a holographic model to solve one of the most persistent puzzles of the central nervous system: Where and how are memories stored? Where and how does perception take place?

Over the years, researchers have spent much time

and effort mapping the brain, showing that specific mental functions take place at specific locations. They have shown that the brain does seem to have its specialized parts: one area for vision, another for hearing, another for certain memories, and so on. But there's a catch. Through accident or surgery, a person loses most or all of one of these parts, and, lo and behold, the memory or ability thought to be entirely localized in that part is still present in the person's mind. It may be fuzzier—resolution may be lost—but the function is there, in toto. Of course, this isn't always the case, but it happens often enough to cast doubt on any theory that insists on specifying the exact location of any mental function.

Then there are more startling discrepancies. Researchers destroy 80 percent or more of the visual cortex of a rat's brain, yet the rat loses none of its ability to respond correctly to visual patterns. Researchers cut away 98 percent of the optic tracts that lead from a cat's eyes to its brain, and still the cat can recognize patterns in the whole. Other researchers combine the two experiments into one simultaneous assault, with little effect on a cat's ability to recognize patterns. Just where in the visual cortex is the function located? Obviously, it's located *everywhere;* every separate fragment of the cortex seems to contain all the information necessary for pattern recognition. Just how was that information encoded in the first place? According to Karl Pribram, through holough holography, through holograms created at the intersecting patterns of electrical activity among brain cells.

"A hologram arises in any system, whether optical, computer or neural," Pribram tells us, "when neighborhood interactions among elements (e.g., spatial frequency) become encoded in the process of transformation."* He points out that holographic encoding is particularly advantageous in the brain. As we've seen, it produces great resistance to damage, since an entire image can be reconstructed from very small pieces of the hologram. It provides a most efficient means of memory storage. (A hundred million bits of information have been stored in a one-millimeter fragment of a physical hologram.) It creates a handy system of cross-referencing and association. "When only part of the input that originally constituted the hologram recurs," Pribram writes, "the remainder of the scene is reconstructed as a *ghost* image."

The holographic model of the brain doesn't deny specialization. Just as light is both particle and wave, so the brain is both specialized and holographic. As in the brain, so in the world. Separate objects and organizations do exist, having their own unique identities. The objects and organizations are also holographic in effect, somehow containing essential and complete information about the whole universe.

This *both/and-ness,* the persistence of seeming contradiction, confronts us wherever we turn in modern physical theory, and serves the twentieth century as a Zen *koan.* When the master asks, "What is the sound of one hand clapping?" he is not expecting a sensible answer but rather hoping to shatter conventional modes of thought and thus bring you closer to the pure, transrational heart of the world.

Quantum theory is the ultimate *koan* of our times. Experimenters find it to be useful, indeed indispensable, in their work. Then they follow its premises to their quantum-logical conclusions and find them-

---

* Quotations and paraphrases in this paragraph from Karl H. Pribram, "How Is It That Sensing So Much, We Can Do So Little," in F. D. Schmatt and F. G. Worden, eds., *The Neurosciences,* III (Cambridge, Mass., 1974).

selves face to face with seeming absurdities. First, as we've seen, there's that bothersome Bell's Theorem, connecting all spatially separated events. But it's even worse than that. For quantum theory really to work, to put it into everyday language, each electron has to "know" what all the other electrons in the universe are doing in order to "know" what it's to do. It's as if at every point in every electromagnetic field there were a tiny supercomputer that was constantly figuring out everything that's going on in the universe.

Some physicists regard this as an insoluble problem. But what is a problem for one physicist is a solution, an illumination, for another. If you could look at the world of matter as a probability-wave hologram, then you could dispense entirely with the idea of tiny supercomputers. A probability wave, is a mathematical formulation that exists in an n-dimensional space —that is, space with as many dimensions as you please. This space, we might say, is similar to that in which the mind operates. According to quantum theory, every particle can be described as a wave function having three dimensions of space and three of momentum— a total of six dimensions in this n-dimensional space. Two particles can be described in twelve dimensions, three in eighteen, and so on. In this way of conceptualizing, every particle is connected, and the whole universe can be described from the "point of view" of a single particle. As bizarre as all this might sound, probability waves are continually being used in real experiments that yield real results.

With this in mind, it is a large jump, but not an impossible one, to conceive the events of the universe as being actually made up of interference patterns of probability waves. If you can come this far, the rest is easy: In such a universe, information about the whole of it is available at its every point. As with the photographic hologram, the information in the smaller fragments may be fuzzy, poorly resolved. But it is there.

The implications of such a formulation are intriguing. To take it back home, you can look to your own body and say that each subatomic particle within you is in touch with all that is. The rhythmic emptiness we have discovered deep within the proton is in some way a "model" of the entire universe—and so are the atoms, the molecules, cells, integrated systems of cells, organs, and organ systems (cardiovascular, nervous, lymphatic, digestive, and so on), and finally the whole being, the living, breathing, perceiving, conceptualizing person.

Up to this point, for the sake of convenience, I've used "hologram" in a general and rather loose way. The term actually specifies the *visual* record of interference patterns, as on a photographic negative. You should be warned that the entities of which I speak here are not holograms in the narrow sense, but rather exist in a mode *analogous* to that of a hologram. To be more precise, I'm now going to introduce the term "holonomic," which means *named for* or *in the nature of* a hologram. "Holoid" will stand for an entity that is holonomic; and "holónomy" (with the accent on the second syllable) will denote the *quality* of being holonomic. Thus, you might think of yourself, not only as a series of rhythmic fields within fields, but also as a series of holoids within holoids. Just as you have identity, you also have holonomy.

This being the case, strange as it might at first seem, the very structure of your body and being may be said to reflect the ongoing structure of the universe. And all possible knowledge, in its general outlines if not in its sharply focused particulars, is potentially available to you, not just knowledge of the present, but of the past and of at least some of the future. (In a later chapter we'll see that the future contains an element of genuine surprise, which increases exponentially with time.)

These implications of quantum theory resonate with the deepest intuition of the ages, the direct experience of the most revered spiritual masters, and the thought of such philosophers as Leibniz and Spinoza and Whitehead. Our own cultural conditioning, for good and sufficient reasons that I have explored elsewhere,* has tended to objectify and particularize our perceptions, and thus has reduced our chances of perceiving any kind of all-in-oneness. But the perception endures, affecting even the most secular Westerner in moments of unexpected illumination, moments often associated with extreme stress or the approach of death—as we'll see in a later chapter. And now the unexpected and ironic conjugation of modern science with this ancient wisdom!

Let's take another look at the remote-viewing experiments at Stanford Research Institute. This time let's assume that the person closeted in the shielded room is (like everybody else) a fragment of the universe, containing in some form all the information it takes to make it up. The information is contained in the person's very structure; or, to put it another way, the person's structure is made up of information from the probability-wave holoid. *Thus, no additional input from any outside signal source is needed.* The information is not sensory in the usual sense. It is structural. The remote viewing process does not involve extrasensory perception (ESP), but rather what might be called metasensory perception (MSP). The question of what senses are used becomes irrelevant. Just as memory

* See my book *The Transformation* (New York 1972), especially Chapters 4 and 5.

requires no sensory input, yet is expressed in sensory metaphor, so it is with this holonomic information.

The target is the Palo Alto Yacht Harbor at low tide. The dark, oozing, viscous mud bottom is entirely exposed between floating docks. The docks are resting directly on the mud. Hal Puthoff is standing on one of the docks looking down at it.

Twenty minutes *prior* to this time, Hella Hammid sits in a room at SRI several miles away and describes the scene as "some kind of congealing tar, or maybe an area of condensed lava. It looks like the whole area is covered with some kind of wrinkled elephant skin that has oozed out to fill up some kind of boundaries where Hal is standing."

She has described the essence of the scene. She has expressed shape, form, color, and material in a manner that allows unmistakable identification. She has not identified the features of the scene by name or expressed its function; that information is cultural in nature, perhaps less than fundamental in terms of the universal holoid.

In another instance, the target is a playground. Puthoff is watching a child swinging on a squeaky swing. Viewing the scene precognitively, Hella Hammid repeats again and again that the main thing at the remote site is "a black iron triangle that the outbound experimenter had somehow walked into or was standing on." She says the triangle is "bigger than a man," and she hears a "squeak, squeak, about once a second."

She has not called the object a swing, but has succeeded in capturing the essence of the scene, including the squeak of the swing. We might say the picture is complete if somewhat fuzzy. The resolution is poor, just as an image reconstructed from a tiny fragment of a photographic hologram would be.

The probability-wave holonomic hypothesis is not necessary to explain remote viewing. It could be explained, as noted earlier, in terms of Schumann waves of extremely low frequency, though this now seems unlikely. Or we could imagine other, as-yet-undetected "psi waves" as carriers of information from mind to mind. But the holonomic hypothesis has a compelling beauty all its own. It stretches us toward the universal and introduces a whole new context of information transfer.

In terms of information theory, memory and imagination would tend to interfere with remote viewing whether or not we use the idea of the hologram. Thus a meditator, one practiced in clearing the mind of this kind of noise, would seem to have an advantage in the skill. Using the holonomic hypothesis, we might expect another factor to enter the equation: the more complex and highly organized the being doing the viewing, the better the resolution of the scene. An amoeba, like a human being, would "contain" the whole picture, but it would not be able to resolve the picture nearly as well as would a human. In the same way, when and if we first meet a more highly evolved being from another planetary system, we would expect that being to "see" events distant in time and space more clearly than we do. To follow this line of reasoning all the way, we could expect each step in evolution, each increase in organized complexity, to result in successively brighter and sharper pictures of the All, right up to what we could call God, who would have consciousness of the fully resolved universal picture.

Meanwhile, the world as it is presses in on us, confronting us with puzzles and contradictions on every side, as vividly illustrated in one facet of the SRI experiments. It has probably already occurred to you that a person skilled in remote viewing would make an ideal international spy. The thought most assuredly has occurred to intelligence organizations in the United States and the Soviet Union. While I have seen remote viewing as possible evidence of the connectedness of all life, the CIA has seen it as a possible weapon in the Cold War. CIA agents have interviewed people known to be skilled in remote viewing. Pat Price, who has since died, participated not only in the experiments I have described, but also in an intelligence-type "Project Scanate," during which he reputedly located and described U.S. and Soviet satellite interrogation stations. Remote-viewing experiments, according to some reports, have been partially supported through organizations funded by government intelligence agencies, and rumors abound that much of Puthoff and Targ's work is now classified, and that their publicly reported work is a cover for more ominous psychic experiments that would influence objects or thoughts at a distance. For at least twenty-five years, the Soviets have reportedly engaged in a serious program in these areas for intelligence purposes. Thus, each side can say with perfect self justification, "I'm only doing it because the other guy is."

How ironic this seems, how paradoxical. Yet every paradox is a chamber of mirrors, reflecting reality at unexpected angles and sometimes pointing the way toward a new reality. In that chamber, the dance of destruction can suddenly appear to us as a potential servant of human evolution. Imagine now all those spy satellites that crisscross the heavens, those electronic monitoring stations athwart the wastes of the world, those reconnaisance planes and ships and submarines, those embassy and consulate listening posts, those ingenious agents in the James Bond image, those patient analysts sifting and evaluating hundreds of thousands of reports, those twinkling computers, those underground war rooms with their marvelous electronic displays—a million grownups playing a deadly children's game. And what is the purpose behind all this activity, this lavish expenditure of human energy? It is simply to *know* the other side fully. The job of intelligence organizations is to uncover a potential enemy's resources and weaponry, its order of battle. But it is even more important to know the deepest, truest intention of its key people, their innermost hearts.

In this enterprise, the globe is newly interconnected, and no one involved can be unaware of the invisible, ever-spreading web that joins us, "friend" and "enemy" alike. This connection, held firm by the threat of nuclear holocaust, cannot be cut, only transformed by the evolution of a consciousness that is at least planetary. Remote viewing will probably fail as a significant spy device—or as a tool for Big Brother—being too capricious, too resistant to the analytical mode. But the mere possibility of a closer connection from one human heart to another cuts through the Cold War spy network and suggests a next step in human evolution: the development of an order of being that can bring the consciousness of all humankind to the *self-aware* consciousness of every human individual.

That information is already present, waiting to be perceived. In fact, as we have seen, it is possible to conceive of each human individual as *consisting* of pure information expressed as rhythmic waves that start as the infinitesimal vibrations of subatomic particles and build outward as ever-widening resonant hierarchies of atoms, molecules, cells, organs, organisms, families, bands, tribes, nations, civilizations, and beyond. At every step along the way, every entity is connected to the great web of information that is the universe. At the most fundamental level, the connection is not sensory but structural, for we are not *in* but *of* the web of relationship. As part of the web, each of us *is* an individual identity, and that identity can be most easily expressed as a wave function, a unique rhythmic pulse. Simultaneously, paradoxically, each of us *is* a holoid of the universe, and that holoid is also expressed in terms of wave functions. Thus, we are both individual and universal, and the web of relationship involves both aspects of our being. Though all the information of the universe is ultimately available "in" each of us, the amount of it we can encode and express—a tiny amount, indeed—is limited by our particular history, culture, language, and nervous system. For now, it is enough to say that we are completely, firmly, absolutely connected with all of existence, and that the next evolutionary step will involve, at the least, our realizing that connection.

SEVEN

The Playground
of Reality

Every theory of the universe or the brain is essentially
a metaphor, in which something unfamiliar is described in terms of something less unfamiliar. The
best we can do in any explanation (no matter how
penetrating our instruments and methods) is to create
metaphoric structures of varying complexity, beauty,
and tensile strength. The connective strands in any
such structure may endure for a while, but must eventually weaken under the strain of new data and ideas,
new language. Explanation is useful, illuminating, in-
escapable. At best, however, it is tentative and provi-
sional.

   With this in mind, I plan at this point to sum up the
explanatory scheme that has been emerging in these
speculations on the subject of rhythm. We begin with
a question: What is a human being? The answer I
propose by no means denies the familiar constructs of
physics, biology, sociology, and the like, but deals with
more fundamental categories. Stripping the question
bare of all but the irreducible, we are left with two

essential qualities of being. These qualities apply in some way to every entity in the universe, but here, for the sake of convenience, I'm making my argument in human terms. Thus:

1. A human being is a "hologram" or *holoid* of the entire universe. We can call this quality *holonomy.*

2. A human being is also an unique entity, expressing a quality I have called *identity,* which is discrete and positional in time and space, and which can also transcend time and space as we know them. This unique identity can be summed up as a distinctive pattern of rhythmic waves or pulsations.

The holonomic formulation was treated at some length in Chapter 6. It resonates with one of the most ancient intuitions of the race, expressed with eloquence and force in Eastern philosophy. It helps account for the essential meaningfulness of existence, the coherent, repeated patterns that we keep discovering at the deepest structure of language, mathematics, and the physical world. It is a necessary consequence of modern quantum theory taken to its logical extreme. And it provides a tentative explanation for certain extraordinary phenomena that have not yet been explained in terms of our conventional reality structure.

This idea of the universe as an overarching unity repeated somehow in each of its parts bears a majesty and elegance all its own. But it leaves something lacking. What of the personal self? How about love, desire, creation, conflict, birth, death, society, history, evolution—the whole world of relationship? Eastern philosophy at its most unequivocal (as in certain forms of Theravada Buddhism and in the Mayavada Vedanta) disposes of all these considerations as *maya,* illusion. This turn of thought has sometimes created a passive otherworldliness that ignores the problems and play of this world while fostering an elitist devotion to satori or *samadhi,* the submersion of the illusory self into the One.

We of the West, of course, have become all too familiar with an opposite imbalance: a preoccupation with time and history, ego and action. This imbalance has its own perils, creating personal anxiety and alienation and a tendency toward fragmentation and exploitation of the natural and human world.

On purely practical, social grounds, then, we would want to avoid exclusive devotion either to holonomy or to identity. But quite apart from social considerations, there is something structurally unsatisfactory about the idea of a human being as *just* a holoid or as *just* a unique self. What I am suggesting is that both qualities, in an equal and complementary relationship, stand at the heart of human existence, and of the existence of the universe.

My proposition is a simple one: If there were no identity—that is, if all the universe were merely the One reflected infinitely in every "part" of the One— then there would be no real *differences,* and thus no relationships would be possible. And without relationship, there could be no object or event, no physical universe, no mental universe, no spiritual universe— no *thing.* To take an extreme view, it is out of relationships that what we call objects emerge; things can be seen as the precipitates of relationships. At the least, we can say, not that relationships create things or that things create relationships, but that the two are mutually dependent, complementary.

A difference that makes a difference, Gregory Bateson tells us, is the elementary unit of information. It is also the fundamental unit of existence, since relationships are possible only between or among entities that are somehow different. Furthermore, for the universe to manifest any kind of stability and persistence (which is to say, "meaningfulness"), the different entities in it must have a certain stability and persistence.

Which takes us back to *identity*, the stable, persistent, unique quality that manifests itself in so many ways in every human individual. We have seen how personal identity, in the form of a distinctive pulse, can persist over time in the works of great classical composers, and I have speculated that the same thing is true in all truly creative personal works. We have noted a number of bodily functions that display identifiable patterns: face, fingerprint, voiceprint, vibrato, breathing pattern, brain wave, DNA molecule. From this, I've speculated that all significant bodily and behavioral functions, when subjected to adequate analysis, will be shown to display unique, identifiable rhythmic patterns, the sum of which can be taken as an overall wave function, the most fundamental quality of individuality.

There is something else—strange, overwhelming, inexplicable—a sense toward which words can only grope: "This is *I*, the whole of existence, around which all the world revolves. Yet how can this be, with you and you, and you also the world and all its works? How did it come to be that in all of space and all of time, I am I?" This powerful sense that seems to strike with the force of a bullet just before sleep or while coming around the bend of a mountain trail obviously has something to do with identity. But identity alone can't account for it. Animals and trees and newborn babies have identity while lacking this sense of *I-ness.*

How is *I-ness* related to holonomy and identity? Is there some commonplace explanation, or is it even stranger than we might have imagined?

At this point, I want to tell the first of two stories that might help illuminate the interplay between holonomy and identity, and show us how the stuff of our everyday life relates to these underlying qualities of existence.

This Isn't Richard

There were four of us, three men and a woman, who shared an ordeal, a rite of passage. In a culture that has turned its eyes from challenge and chance and possible tragedy, this was a rare gift indeed. We were up for black belt in aikido—Richard, Lawrence, Wendy, and I—and over a period of about a year each of us in our own separate ways confronted injury, exhaustion, humiliation, and despair. Our list of injuries alone suggested the severity of the ordeal. In addition to numerous bumps, bruises, and abrasions, we suffered a broken foot (Lawrence), a sprained neck and torn ligaments of the elbow (Wendy), a fracture of the cheekbone and a multiple fracture of the arm (Richard), and a dislocated shoulder (me). These injuries might seem excessive in an art that so often has the effortless quality of a dance or a dream. But in aikido no punches are pulled, and each attack proceeds to its logical conclusion, with the attacker pinned or thrown through the air. Thus, the aikidoist must practice hard and long to transform the fear of falling into the joy of flying—an unforgiving if ecstatic practice.

But we should not linger over injuries, for that would only distract us from the true significance of what we faced. Our teacher, Robert Nadeau, is not your run-of-the-mill martial artist. Though skilled in judo and karate as well as aikido (and, we all agree, a formidable man to meet in a dark alley), he views himself as primarily a teacher of meditation and alternative ways of handling life's pressures. For Nadeau, the mat is the world. Thus, he teaches us not to deny or avoid tensions and problems and pain in our practice but to welcome them as treasured gifts, as opportunities for transforming our lives. Far from working around our weaknesses, Nadeau zeros in on every divided motive, every pretense, every secret, well-guarded flaw. If the mat is the world, it is the world beneath a magnifying glass, where nothing can remain long hidden. In this setting, self-examination is not mandatory; it is unavoidable.

All of this comes to a painfully sharp focus during the three-month period leading up to the black-belt examination. Nadeau uses this period not only as an intensive cram course in advanced techniques, but also as a physical and psychological trial by fire. Anywhere from three and a half to ten years of practice might precede this ordeal. During this period, the candidate is expected to attend all of Nadeau's classes. For the first hour of training each night, which is devoted to basics, the candidate or candidates practice along with all the other students. When the second hour, devoted to advanced training, begins, Nadeau sends the candidates to the back mat. While the other advanced students go on with their regular training, the candidates practice the specific techniques that might be expected during the exam.

Fifteen minutes before the end of class, Nadeau seats the advanced students in the traditional Japanese meditation position around the edges of the main mat and calls the candidates front and center. He puts them through their paces, one by one, as the others watch. At the end, Nadeau arranges for a series of multiple attacks. First one, then two, then three, and even up to seven advanced students are directed to attack the candidate again and again. This goes on until the candidate is reduced to total exhaustion and either trapped or felled by the attackers.

Nadeau has a remarkable ability to know exactly in what technique each candidate is unprepared. On one occasion during my own ordeal, he told me to spend the entire advanced class practicing a technique called *irimi-nage* (entering throw). After forty-five minutes on the back mat, I was called front and center with the full expectation of demonstrating my well-practiced *irimi-nage*. As I stood there waiting for the attack, however, I made the mistake of saying to myself, "I hope he doesn't ask me for *koshi-nage* [waist-throw]." As if in response to my unspoken words, Nadeau said, "Okay, George, let's see your *koshi-nage,*" after which he let me flub one waist-throw after another until everyone present was painfully aware of my unpreparedness.

Over a period of nine months, Lawrence, then I, then Wendy faced our separate ordeals. Under physical and psychological attack, we discovered that a flaw is corrected only by being revealed, and that the true opponent is the one who resides within. Each of the three ordeals contained the tension and danger, the dark despair, the ironic twist, and the happy completion of which our most ancient and cherished tales are made. But it was Richard's experience that seemed to draw us into other worlds, joining us with the immense and the infinite.

In his early thirties, at the very prime and glow of life, Richard might have been a figure from the Elgin Marbles. With his finely muscled, perfectly balanced body and handsome face, he seemed a modern counterpart of the classic Greek ideal of physical beauty, and indeed he had been an Olympic athlete in his college days. Nor were Richard's gifts merely physical. He held a Ph.D. in psychology and was co-founder of a respected school of meditation, body work, diet, and interpersonal relations. He was a superb aikidoist.

Dazzled by his gifts and grace, we might find it hard to discover any flaws in this man, and sometimes Richard did seem almost too good to be true. But eventually a certain quality of calculation emerged, summed up in a phrase from gestalt therapy often used by Richard himself: "Taking care of myself." The phrase was not meant to imply selfishness, but simply to clarify the healthy, openly expressed self-interest that can save you from dependency and the victim's role. Still, you couldn't help noticing that Richard wouldn't accept any invitation until he had carefully calculated what he would get out of it; then he would accept only if, on balance, he figured he would come out on the plus side. Richard rarely did anything on speculation. He took good care of himself.

There was, as well, the matter of name. Richard had contributed a great deal to the field of human growth. Yet a number of other people who had made lesser contributions had become better known. Richard was aware of this and, it seemed to me, had a burning if rarely expressed desire to make a name for himself.

I doubt very much if our teacher involved himself in this sort of analysis. He simply intuited, then acted: When he had told Wendy she would be going up for her test in three months, he had told Richard that he didn't know whether he would be going up or not. Richard could go through the three months of preparation if he wished. On the day of the exam, said Nadeau, he would let him know whether or not he would take it.

For Richard, this was like a slap in the face. He would have to endure a three-month-long ordeal with no assurance of any reward at the end. When fellow students would ask him if he was going up for black belt, he would have to say—though he would obviously be practicing hard for the event—that he didn't know. Not a very good plan for someone used to taking care of himself. Already, he had suffered the most serious injuries of anyone in the school. After breaking his arm (during a strenuous throw), he had continued practicing while wearing a cast. Later a break in his cheekbone (he had been kicked accidentally while down) had temporarily affected an eye muscle, and still he had continued practicing. Surely he had paid his dues in full.

Yet there he was on the back mat night after night, driving himself to exhaustion in the face of uncertain odds. As the weeks passed, Nadeau paid less and less attention to him. One night two weeks before the exam, I happened to be sitting next to Richard at the edge of the main mat at the end of class as Nadeau put the candidates through their paces.

"Is there anyone else?" Nadeau said, looking right past Richard.

Richard said nothing, and I heard myself answering for him. "There's Richard here. You forgot him."

"Oh, yeah," Nadeau said dryly. "What's-his-name. Okay, let's see what he's got."

From this moment until after the exam, Nadeau never looked at Richard or called him by name. Occasionally, he would summon "what's-his-name" to the center mat, and then make no comment about his performance. In the shower room three days before the exam, I asked Richard what he thought was going on.

"I don't know. I can't tell exactly. Something's happening to me. I'm beginning to feel some kind of transformation."

As is the custom, the exam was scheduled on a Sunday at one. It was a beautiful, cloudless June day. People began gathering early: aikidoists from miles around, hundreds of spectators. An examining board of five ranking black belts would be convened to pass on the candidates' performances.

"Well, are you going up?" I asked Richard when he appeared on the mat.

"I don't know. Nadeau still won't speak to me."

The *dojo* had the feeling of a church before a wedding. Some people were meditating. Others were talking in hushed tones. Richard went into the office and came out with a strange look on his face.

"I guess I'm going to take it," he said. "I saw my name on the schedule. Nadeau still hasn't said anything."

To begin the ceremony, all the aikidoists bowed to the portrait of a venerable Japanese warrior on the front wall. This was Morihei Uyeshiba, the legendary founder of aikido, known to all as O Sensei, the greatest of teachers, whose seemingly miraculous feats in his old age had been photographed, filmed, and confirmed by respected witnesses. Five candidates had already been examined when Richard was called to the center of the mat. With his *uke* (oo-kay, "attacker") he moved out in the graceful knee-walk common to the art. The two of them bowed first to O Sensei, then to the five examiners, then to each other. Nadeau called out the first series of techniques, and the exam began.

From the very beginning, it was apparent that something extraordinary was occurring. It was like one of those sporting events that are later memorialized, perhaps a World Series game or a bullfight, during which every last spectator realizes at some level that what is happening out on the field is more than a game, but rather something achingly beautiful and inevitable, an enactment in space and time of how the universe works, how things are. As Richard and his *uke*, still on their knees, glided through a series of attacks and pins as precise and formal as a tea ceremony, the silence in the *dojo* became deeper and more vibrant. Nadeau called for the next series of techniques. The *uke* rose and attacked the still-kneeling Richard, who moved in sweeping circular motions to embrace the attack. So gentle and coherent were his movements that they seemed to capture time itself and slow it to a more stately pace. Sometimes when Richard pinned his attacker with one hand, he reached out with the other in a gesture of balance that I had never seen him use in practice. This supple, rather androgynous movement was obviously not needed for balancing the physical body. It was as if Richard's hand were reaching beyond the four walls of the *dojo* to a point of balance in the cosmos.

Nadeau called for the next series of techniques, which would have both attacker and defender standing. When Richard rose to his feet, there was a slight stir in the room; people here and there glanced up at the windows or the lights. What had happened, inexplicably, was that the room had suddenly become appreciably lighter.

From this point to the end of the story, I am relying, not just on my perceptions, but those of several other people, including Richard, all of whom I phoned the next morning. Without telling any of them what the others had said, I began to piece together a coherent account of the previous afternoon's events. My informants did not agree in every particular, but there was more agreement than disagreement, and a clear general picture emerged. I present it here simply as a consensus of subjective reports.

97

Everyone I contacted noticed the shift of illumination when Richard rose to a standing position. Some people also began seeing an aura—some described it as "golden," others as "clear plastic"—around his entire body. As the exam continued, the speed and intensity of the attacks increased, and yet there was still a general sense of time's moving slowly, at an unhurried, dreamlike pace. The spacious *dojo* began to seem smaller; an unfamiliar feeling of intimacy came over the aikidoists and spectators around the mat, as if we were involved together in something usually reserved for our most private moments. During one swift attack, a hard strike to the belly, Richard slipped quickly to the side and made a bewildering gesture that none of us had previously seen. The *uke*, without having been touched, went down with a loud crash. This rather formal young man, a stickler for decorum, lay there for a moment looking up at Richard in astonishment, then laughed aloud. Later, Richard could not recall or reconstruct this remarkable technique. For his part, Richard was beginning to get the feeling that he was not "doing" anything at all, that the movements of his body were "just happening" without thought or effort.

The exam continued in this spirit, like a long, hypnotic phrase of music, through the body throws and defense against knife. Then, when Nadeau called for the *uke* to attack free-style, the illumination in the room seemed to go up another notch and the boundary of light surrounding Richard seemed to become denser, brighter, and unmistakably golden. The genius of aikido is to transform the most violent attack, by embracing it, into a dance, and it was the essence of dance we saw there on the mat—neither powerful nor delicate, neither destructive nor creative, neither masculine nor feminine, but all such seeming opposites connected and drawn to a point of balance.

At a particularly radiant moment, Nadeau stopped the free-style attacks and gave Richard a minute to catch his breath before the climactic *randori,* the multiple attack. Richard turned away from the audience, in accordance with *dojo* etiquette, to straighten his *gi* uniform. As he did so, he glanced up at the portrait of O Sensei. A powerful arc of golden light seemed to be streaming from the eyebrows on the picture toward Richard's head, covering him, suffusing him with gold. At this moment, we in the *dojo* experienced a third brightening of the room. By the time the three-man attack was in full swing, the whole place was alight as if from within with the most delicious, joyful, almost palpable illumination.

To a first-time spectator, the rushing, swirling, tumbling, crashing motion of a *randori* is simply overwhelming; the senses can't handle it. An expert aikidoist observes techniques and moves, watches for breaks in the energy field that subsumes both defender and attackers. But on this day spectators and experts alike saw Richard's *randori* as harmony, the promise of reconciliation. No matter how hard or swift the blow, he was not there to receive it, but always at the moving center that holds all opposites in perfect tension. As for Richard, he experienced no effort or strain whatever; only a voice in his head, repeating, *"This isn't Richard. This isn't Richard."* There, in the eye of the storm, stripped of the certainty he had always deemed necessary for survival, denied the support of his teacher, divested even of his name, Richard found the deliverance he had not known he was searching for. He had no question that he would be hit or trapped. If need be, he could go on forever, realizing all the while that "he" was not doing it. The voice in his head was clear: *"This isn't Richard. This isn't Richard."*

"After the *randori,* " one of the people I called the following day told me, "I just sat there stunned. I couldn't even move. It was only when the next exam started and the guy's technique was so crude compared to Richard's that I was shaken to the realization of what was going on—and that was *really* awesome. *O Sensei was in that room. I knew* it. I could feel the presence. Those crude techniques gave me the contrast I needed to sense it. O Sensei had been there all during Richard's exam."

EGO AND IDENTITY

The boundaries of what is "real" are not nearly so clear these days as they once seemed. A number of observers have pointed out that perceptions are not purely objective, but rather are strongly influenced by a social consensus, which is impressed upon us subtly and not so subtly from earliest childhood. Consensus reality, we have been told, is determined to some extent by custom and convention. It is also strongly influenced by the range and the very structure of language. Experiences and perceptions for which there are no commonly accepted words are often invalidated and sometimes entirely extinguished.*

Consensus reality is shaped and guarded, as well, by a pervasive constraint that goes under the general term "ego."† Ego is not a structure *within* the brain, body, and senses, but rather a particular way of structuring the relationships among brain, body, senses, and environment. Ego ensures that the individual apprehends himself or herself as a separate, self-aware, volitional entity who acts upon the world and is acted upon by the world. Through this ego apprehension, the individual can experience inflation and guilt, can take credit and blame. Moreover, the ego not only reflects but is *made of* the various stratagems—most of them, as Freud points out, derived from early family interactions—that the individual has developed for dealing with the world.

In our culture, the first faint outlines of ego begin appearing at about age two. Then, like an image on a developing photographic print, it gradually takes on

* See my article "Language and Reality," *Harpers,* (October 1974).

† "Ego" is used here as a basic reality constraint, and thus subsumes the Freudian ego and superego.

form and depth and detail in the pervasive bath of social conditioning until, by age eight or nine, the picture is fairly complete. Each culture tends to produce certain ego characteristics. Thus, the Western ego is said to be more oriented to time and progress than is that of the East, which focuses on the spatial aspects of existence and takes a relatively serene and cynical view of progress. In any case, the normal individual in all modern cultures does develop an ego, and civilized societies as we now know them would be impossible without it.

Beware, then, any casual denigration of the ego. A person who fails to develop a strong ego is not only adrift within society, but also is a poor risk for those journeys of the psyche that press at the boundaries of the social agreement. At the same time, it's clear in the teachings of all the great spiritual masters that a deeper and more vivid reality lies beyond language, beyond social convention, and especially beyond ego. Throughout history, these masters have devised rigorous and ingenious methods for weaning the spiritual initiate from his or her lifelong reality structure: various monastic disciplines, the verbal mystification of Sufi stories and Zen koans, and direct attacks on language and ego.

Perhaps you have already recognized Robert Nadeau's treatment of Richard as a classic example of the process known as "reduction of the ego." We met Richard as a gifted, capable man with a strong, rather complex ego. Three months later, after seemingly cruel and humiliating treatment, we saw him on the mat, stripped of the very aspects of his ego that he considered most important to his success and survival —stripped at last, if only briefly, of ego itself. We then witnessed what any knowledgeable observer would

call an exceptional, artistically transcendent performance. We discovered, moreover, that Richard's experience apparently affected or was shared by many of
those present in a synchronous and generally consistent manner. I want to take this as a case in point, not
as objective fact but as pure experience, and discuss
briefly how it relates to what I have called identity and
holonomy.

Ego can be seen, first of all, as a specific, unequivocal constraint against the experience of holonomy. It
allows perceptions of separateness from the world and
disallows perceptions of oneness with the world. The
central statement of Eastern mysticism, *tat tvam asi*
("Thou art That," meaning that the individual is as
one with everything else), presents a challenge to the
ego of any culture and constitutes an offense to the
ego of the West. To achieve the merging with the One
that is the sine qua non of the mystical sensibility, it
is surely necessary to get rid of ego. Always on guard,
ego rules out the experience of oneness and makes the
experience of connectedness difficult and unlikely.
The "I" of ego is a doggedly separate "I."

Thus we might think that this ego, being so anti-
thetical to holonomy, is invariably allied with or even
the same as the essential human quality of identity. I'd
like to propose, to the contrary, that ego can constrain
against the experience of identity just as it constrains
against the experience of holonomy. Ego judges; identity is beyond judgment. Ego either anticipates or observes the eternal moment from which the universe
unfolds. Identity exists only in that moment. Ego is an
entity summoned up by society to serve both society
and the individual at the dialectical point of tension
between the two. Identity serves nothing; it simply *is*.
Ego springs from the distinctive rhythm of identity
but, to the extent it is swayed by the contrary dictates

of society, can suffer disrhythmia. Identity is always in perfect rhythm, for it *is* that rhythm. Ego can adopt a name, and a name can have deep roots that reach down toward the domain of identity. But identity itself has no name, for a name can be duplicated, and the essential rhythm of identity is unique among that of all other entities in the universe. *"This isn't Richard,"* then, was by no means a lament over the loss of identity but a cry of joy over its discovery.

Ego can constrain against the experience of identity as well as against the experience of holonomy because, in essence, identity and holonomy are two aspects of, two ways of looking at, the same thing. Holonomy is the immanent unity of the universe, which is somehow reflected or contained in its every entity. Identity is the multiplicity of the universe expressed as the uniqueness of its every entity.

The ultimate experience that lies beneath custom, language, and ego is our apprehension of the interplay between the two. Human experience binds and is bound by this interplay, which is to be found only in the intensely burning, never-to-be-recaptured present moment, and in which the mysterious, overwhelming sense of "I-ness" is to be found. Here, at the point of interplay between seeming opposites, all is present and perfectly connected, nothing is impossible, and there is only harmony.

Our next story shows what can happen at that point. Touching as it does upon the fierce, uncertain yoga of love, this story resists simple exposition. The novelist, V. S. Naipaul, has written: "An autobiography can distort; facts can be realigned. But fiction never lies; it reveals the writer totally." Let us, then, seeking absolute truth, change the characters' names, and set the following events in the revealing idiom of fiction.

A Lone Biplane

Bill was a writer of sorts, Mary an artist to the hilt. They met casually, their acquaintance developed into friendship, and their friendship took about two minutes to turn into attraction, after which it was only a matter of time before they would become lovers. On a fine April day, they had a light lunch and a bottle of wine at a restaurant overlooking the sea, then took a short walk to a nearby amusement park before driving to the lonelier stretch of beach where they had reserved a motel room. With every step, the constraints of their customary lives fell away.

"Do you want to take a ride on the roller coaster?" Bill asked.

"I don't need to," Mary said, holding her stomach. "I'm already on a roller coaster."

"Me too. To tell the truth, I hate those things. It's funny —in the Air Force I used to do every kind of acrobatic you can imagine. I could hang upside down in an open-cockpit biplane and feel perfectly happy. But roller coasters scare the hell out of me. It's different somehow."

With these words, a powerful memory possessed Bill, and he began telling Mary about his days in Air Force training, about the Stearman biplane in which he and thousands of other pilots had learned to fly, and about the undiluted joy of climbing to 5,000 feet on a hot summer's day and doing rolls and loops and wingovers all the way down to 1,000 feet.

"Do you know what a snap roll is?" he asked.

"I haven't the faintest idea," Mary laughed. "What is a snap roll?"

"Well . . ." He paused to look in her eyes, then went on. "It's like this: You dive a bit to pick up speed, then you pull back on the stick and when your nose is just above the horizon, you really haul back on the stick, like this, to create a high-speed stall, and you whip the stick all the way over to the right and at the same time push in full right rudder. And then it's like everything comes unhinged and the wind whistles through the wires and struts and you spin right over on your back, and when you're exactly upside down, you give it full left rudder to slow the rotation, and then, when you're just about level again, you pop the stick forward and neutralize the rudder. Am I being clear?"

"Totally confusing," she laughed. "Please go on."

"I really do wish there was some way I could show you. I wish I had a Stearman. There were thousands of them at the end of the war and you could get them for practically nothing. Now you can't get one for any price. God, I wish I could show you."

For some reason, it did seem terribly important to Bill. There was a tingling at the back of his neck, as if this matter of the Stearman and the snap roll had some deep significance.

"How about a loop?" he said. "Surely you know what a loop is."

"No, I don't," Mary said, casting a smile his way. "Tell me what a loop is."

He proceeded to explain, knowing that the literal meaning of his words didn't matter and that whatever he said and whatever she said would be all right. But there was still that tingling at the back of his neck and a slightly eerie sense that the subject he had chanced upon had a meaning that was beyond his ability to comprehend.

By the time they were in their room, however, the subject was forgotten, and they came together with a minimum of hesitation and self-consciousness. It was a validation of every look and word and touch that had passed between them and, more than that, a stripping away of every remaining constraint. And there was something else: At one point, they passed beyond ego, even beyond desire, and entered that state in which there is only a sense of existence at its most primal and vibrant. It is perhaps toward that state, in spite of all the tortuous detours we have invented, that all sexual love leads. To enter it in this self-conscious culture is rare, however, and Bill and Mary reemerged into the world of ego and desire somewhat shaken.

They dressed and went out for a walk on the beach. A

cover of high clouds, light grey with a faint mother-of-pearl glaze, had slid over the enormous dome of the sky. There was no wind at all and the sea was exceptionally calm. Soft, sea-fresh air caressed their cheeks, and the sand was cool and moist on their bare feet. Bill and Mary walked close together, not speaking, pervaded and surrounded by an unfamiliar sense of clarity and quiet. There was no hurry now; time was of no concern to them. They passed only one person on their walk: a young woman carrying a bucket and a spade. Her face was round and rosy and she smiled at them as if she had known them always.

After walking perhaps two miles on this strangely deserted beach, Bill and Mary stopped and looked at each other without pretense or calculation in the even light of that cloud-covered afternoon.

"Shall we head back?" Bill asked.

Mary nodded. As they turned, Bill's eyes were caught by a movement in the sky over the hills just inland. He froze and watched a lone biplane coming from the land toward the sea, straight for the two of them, in fact.

"My God, Mary. I think it's a Stearman. The plane I was telling you about."

The plane continued on a beeline toward them, moving quite swiftly, as if with a definite sense of purpose.

"It *is* a Stearman," Bill said. "It's a jet-black Stearman."

The plane passed directly over them. It was not very high; Bill guessed around 2,500 feet. It continued out to sea for just a moment, then turned sharply and passed over them again. Back over the land now, the black Stearman made a sharp turn to the right, then to the left.

"I think it's clearing the area for acrobatics," Bill said. "That's what we used to do before acrobatics. Maybe it will do a snap roll."

As soon as he said those words, the biplane dove to pick up speed. During the brief dive, the nose of the plane was pointed directly toward Bill and Mary as if very specifically signaling to them; it was not a degree off one way or the other. Then it pulled sharply out of the dive and performed a perfect snap roll.

"That's a snap roll," Bill said. "That's exactly what I was trying to explain to you. My God, this is strange."

The biplane proceeded out over the sea and did two more clearing turns.

"Now it's going to do another snap roll," Bill said quietly.

Again the biplane pointed directly toward the couple, this time from the direction of the sea, and did a snap roll. Bill surveyed his physical surroundings: Mary's eyes, the vast, calm seascape, the immense dome of clouds. Then he looked up at the biplane again. It was not that the physical world was dissolving in his sight. Everything was sharp and clear, fixed inexorably in its proper place, even more immediate and vivid than before. But a momentous change was taking place concerning the connection between himself and all that he perceived. The biplane was turning right and left again.

"Now it's going to do a loop," Bill said, almost in a whisper.

This time the plane didn't point directly toward them, but performed a loop precisely on a line with the edge of the beach. As it pulled up and over, Bill could see a glint of iridescent color on the top of the wings. He realized then that the biplane was not jet black, but rather a deep indigo. Its loop finished, the plane turned in the direction from which it had come and, in the same purposeful manner that had marked its arrival, flew directly away from them and disappeared over the hills.

As Bill stood there looking over the hills where the biplane had disappeared, his mind raced from one possibility to another, seeking some rational explanation for the events that had just occurred. How could the plane have come precisely to their location at precisely the right time and then have performed precisely the manuevers that had earlier seemed so important to him to communicate—and just those and nothing else? It was as if he had been controlling the plane, or had created the entire physical event—or as if the pilot had *known* what he wanted and had proceeded to do it. Could the person in the plane have overheard him talking to Mary earlier near the amusement park, and could that have put the idea of the snap roll and loop in that person's head? But how could that account for the location, which was several miles from the amusement park, and the timing? And, anyway, no one had been anywhere near to overhear him talking.

Maybe it was simply telepathy, with Bill somehow signaling the pilot as to what he wanted. But that didn't make much sense. A more likely explanation, if he had to go to the paranormal, was precognition, with what he had described to Mary in their earlier conversation being based on a vision of the future event. That made at least some sense. Or perhaps, in the dazzling unfamiliarity of the new love affair, he had just imagined or dreamed the earlier conversation, after the fact. This possibility was one he could check out.

"Mary," he said, "did I talk to you about wanting to show you a snap roll and a loop? I mean, did I describe the Stearman biplane and the snap roll and the loop when we were walking right after lunch, when we were talking about riding the roller coaster?"

"You certainly did," she laughed. "More than I could possibly understand about the snap roll."

Maybe it was just coincidence, Bill thought. But, no, the odds were too high; that explanation was unacceptable. And the particular way the events had occurred—the position of the plane, the sequence—could never be reasonably attributed to chance. Bill's mind was spinning. Resisting the deeper truth of the matter—a truth that would require a change in his way of perceiving reality—was disorienting him, making him dizzy.

But there was yet another possibility: Maybe what he had just witnessed had been illusory. Maybe it hadn't happened, or had happened differently from the way he had seen it. But these objections, he thought, were becoming petty. He looked at Mary for some kind of understanding. He knew she was less interested than he in the specifics of the maneuvers, but no less involved in the larger event, the unexpected opening to a new world. Now, as they looked at each other, conventional time seemed to stop and both of them

experienced a flashback to the extraordinary state they had
entered earlier, and a simultaneous flash forward to events
of the approaching evening. They took hands and started
back to the motel.

A couple of hundred yards from where they had stood
watching the plane, Bill and Mary came across another couple sitting on the beach. Taking one more chance to corrob-
orate what he had just witnessed, Bill approached them and,
without any preliminaries, opened a conversation.

"Did you see that biplane a minute ago?"

"Yeah," the man said pleasantly. He was slim and wiry
with dark, wavy hair. The woman was brunette with shining
eyes and a flushed face; she was wearing a lavender smock.
They were holding hands and sitting in the intimate way of
lovers.

"Would you mind telling me what it did?" Bill said.

"It did a couple of rolls and a loop."

"Are you a pilot?"

"I was a radar navigator in B-47s during the Korean
War."

"And you're sure that plane did two rolls and a loop.
What kind of plane was it?"

"Looked like an old Stearman."

"What color was it?"

"I couldn't tell. Looked black."

During this interchange, the man answered without any
sense that being questioned in this manner by a stranger
was anything unusual. In fact, both he and Bill were grinning slightly the whole time they were talking, as if they
were in on some kind of conspiracy. They chatted a minute
or two longer about Air Force experiences before parting,
during which time the two men and the two women seemed
linked in perfect complicity.

After that, Bill simply let go. No longer questioning his
own experience, he lived for a while in a different world.
The joining of the lovers' bodies could as well have been
wind passing through marsh grass or salt dissolving in
water. No effort was required and everything that happened
was interchangeable, inevitable, and of great value. Bill was

neither manipulator nor observer, but rather a glowing presence, a perceiver at the nexus of all events. What he perceived took place. His intentions were the world's intentions. Allowing everyone else the same privilege, he stood at ease at the precise center of the universe. On the golden day he and Mary drove back home from the resort, in fact, he sat at the wheel of the car with the speedometer registering sixty, yet felt as if he were absolutely motionless. It was the city with all its people and buildings, its wonders and delights, that was moving majestically and inexorably toward him, just as he wished it to, just as it was intended to.

As long as he continued perceiving the world in this new way, the world responded with one little miracle after another. Several times, he and Mary met in the most delightful and unexpected manner, through what would have normally been called accidental circumstances, except that now nothing seemed to happen by chance. Driving in heavy traffic became sheer joy, since an opening would always appear for him, sometimes at the last instant. On one occasion, a research paper with data he had been seeking for months appeared mysteriously on his desk; he never learned how it got there. During this period, Bill came to view the miraculous simply as "the way things really are." If explanation be needed, the Jungian notion of synchronicity might serve. And there are more traditional ways of talking about such things: Bill was in a state of grace. He walked with God.

But the forces of homeostasis are powerful, and the world that we have made has a thousand subtle ways of drawing us back from grace. Within two weeks, Bill had lost much of the glow, the effortless certainty that had been given him at the beach. Once again, he perceived himself as separate from the world. He perceived chance events. He perceived effort. As he perceived, so it became.

And yet, something was left from that other world. Never again could Bill pass careless judgment on others' experiences. Nor could he take a single, unyielding view of the nature of reality. Behind the most ordinary event lay the possibility of transcendence. Within the most commonplace object were hidden the varied splendors of all time and form. And sometimes, when his life seemed most fragmented, wearying, and futile, Bill would call upon a memory of the biplane at the beach, the tingling sense of connectedness, the easy miracles of the nights and days that followed. Even in the worst of times, that memory would sustain him.

THE ULTIMATE INTERPLAY

Let us now consider the familiar sensory picture we have of the world around us. When we stand on a hill and survey a landscape, our eyes can focus on only a tiny arc of the entire scene at any given time, yet the whole scene seems to be there all of the time, objective and complete. Misled by the idea of the camera, we commonly assume that there is a picture in our heads that is a reproduction of an identical outside reality. This is by no means the case. The "picture" as such does not exist in the outside world, but rather various frequencies of vibrations as translated through the senses and processed through a series of transformations in the brain, then projected to the "outside." The sensory picture may be conceptualized as a special case within identity—that aspect of it which is discrete and positional in time and space.

There's a problem here. The existence of anything, as we've seen, requires context, relationship. In the conventional view, the sensory picture is a function of the relationship between "inner" and "outer," or between self and other. But if the sensory picture is a special case within identity, then the distinctions between inner and outer, self and other, become less than fundamental. As I've suggested earlier, the ultimate relationship is between holonomy and identity. This relationship is not between "inner" and "outer." It takes place entirely in terms of the context that is a human individual. The picture of the landscape takes its meaning—indeed, takes its existence—from a relationship. This relationship consists of a comparison between the structure of the whole universe as expressed in the context of the individual and a particular, unique point of view of the universe (which includes the sensory picture) as expressed in the context of the individual.

Does this mean that the external universe is unreal? Not at all. When you look at an individual as a context, then galaxies can be billions of miles away in terms of conventional space/time while secret memories are "locked inside." Yet galaxies and memories alike can be seen to exist in the context that is a human individual. You need go no further to discover the playground of reality.

Some 300 years ago, the philosopher Leibniz proposed a somewhat similar idea, with the universe composed of windowless "monads," each of which was a context of the universe from a particular point of view. Leibniz left the work of coordinating the numberless monads to God. Recent insights from the field of holography, along with the research of such neuroscientists as Karl Pribram and such physicists as David Bohm, are coming ever closer to explaining in appropriate modern terms how Leibniz' seemingly mystical monads might work.*

Still, language, convention, and ego conspire to make us think of most relationships as between the inner and the outer, between the self and the other. This conspiracy tends toward the fabrication of a

---

* See Karl H. Pribram, *Languages of the Brain* (Englewood Cliffs, N.J., 1971); and David Bohm, "Quantum Theory as an Indication of a New Order in Physics, Part B: Implicate and Explicate Order in Physical Law," *Foundations of Physics* 3:2 (1973), pp. 139–68.

fixed, objective outer world, which at this stage in human development seems necessary for social consensus and stability. Unhappily, it also creates the experience of opposition and struggle, expectation and anxiety, with the "inside" self experienced as frequently at war with the "outside" environment. At worst, this makes victims or oppressors of us all.

Yet a faint awareness of the ultimate interplay lingers always at the edges of our consciousness, expressed in the mysterious sense of "I-ness" of which I have spoken and, indeed, in our every intimation of a larger destiny. I must stress that this "I-ness" is not mere subjective consciousness. I might best describe it as the ego's wonder at a deeper truth. That truth is simply that the individual, the "I," *is* the universe.

Ordinary reality can be defined as a fortification against paradox. When we venture out beyond this bulwark and approach any one of the great questions —the nature of light, of time or space or consciousness—we meet paradox on every side. In an expanding universe, for example, our own galaxy seems exactly at the center—*and so does every other galaxy.* Every point in this expanding universe, in fact, acts as if it were at the very center of the universe. This might sound impossible, but the mathematics on the subject is quite clear.

In aikido, we are taught that when we become fully aware of and take responsibility for our own center, then our center becomes one with the center of the universe. Does this tend to make us domineering, self-centered? Far from it. Realizing that I am the center of the universe makes it easy for me to realize that you are too, and that there are space and time enough for you and me and all other beings. It is likely to be the person without awareness of a center who is belligerent, acquisitive, selfish—

one who pushes for dominance or mere survival.

This physically expanding universe is, as will be discussed in a later chapter, also a universe of expanding possibilities, of evolution and transformation. Single-celled organisms evolve and galaxies evolve. Species and cultures evolve. And the predominant direction of this evolution is toward increasing complexity and order, with new information, new options manifesting themselves at every point. In such a universe, it would not seem entirely unreasonable to entertain the paradoxical notion that Leibniz solved by bringing God out of the machinery: that is, that each being *is* the entire universe from a unique point of view, and that it's possible for all these universes to work together harmoniously. A sense of such a harmony, in any case, is precisely the knowledge we bring back from those moments when we somehow slip our bonds, let go the constraints of language, convention, and ego, and experience directly the ultimate interplay, the paradoxical relationship between holonomy and identity within the context that we call "ourself."

During those times, as we've seen in our stories, what we intend happens and what happens is what we intend. There is no waiting, since everything is already taking place. There is no unfulfilled desire, since desire itself dissolves on the ever-present instant of fulfillment. There are no chance events, since we are the architects of creation and all things are connected through us. Others around us (other universes) report that they, too, are affected. Miracles seem to happen, eventually seeming less like miracles than "just the way things are." The ego is not destroyed or distressed, as in schizophrenia, but transcended. Each of us, in this ultimate interplay, is like a god, omnipotent and omniscient. Does this mean we can fly or move mountains? It's impossible to say. But whatever happens is just what is intended. In the playground of reality, where the relationship between identity and holonomy is directly experienced, each of us is in the business of creating all of existence, effortlessly, on the wink of an instant.

There's one more thing: The relationship between identity and holonomy, whether we are experiencing it or not, is always going on, a silent pulse at the heart of our experience. This relationship operates—it must operate—in perfect rhythm. However out of joint our affairs, however disrhythmic our experience, during times of suffering and on the approach of death, even in the midst of the most boring routine, it is always there. We have tended to look for this sort of perfection in the esoteric and the mystical. But perhaps, as we'll see in the next chapter, it's more a part of what we call "ordinary" than we think.

EIGHT
Perfect Rhythm

From earliest childhood, we are taught that goals are more important than experience. We learn to distrust our deepest feelings, to override our natural rhythms. Arrogant with our deeds and discontents, we consider it presumptuous to claim even a moment in which our personal rhythms are joined in perfect synchrony with the rhythms of the cosmos. And so we turn to the lives of saints and yogis, those who through grace or life-long devotional practice have managed to strip away ego and desire, who have at last achieved what T. S. Eliot calls "A condition of complete simplicity, / Costing not less than everything."

Yet, at the deepest level, each of us is a yogi and a saint. The relationship between holonomy and identity pulses silently at the heart of every life. Ignore it as we will, it is there to serve us during our times of greatest stress or joy, and sometimes to rise to consciousness when we least expect it. The profane as well as the sacred can lead us to the wellspring of perfect rhythm. The flight of a well-hit golf ball as well as the tones of a Gregorian chant can join us with the music of the spheres.

The moments of perfect rhythm, in fact, are probably far more prevalent than we might imagine. Every one of us, if we are willing to open our memories wide, can bring to mind one or more of these periods when everything seemed "extremely easy" or "exactly right." Words tend to betray these moments, becoming too easily maudlin or pretentious. So we are generally satisfied to say simply, "It's like being just exactly in the right place at the right time." Or "It's as if nothing can possibly go wrong." Or "It's like ice skating. If you go right, that's right, and if you go left, that's right. Everything is easy."

People who are known as "winners," according to one newspaper feature,* are precisely those who frequently reach the state of rightness and ease. Reporter Jerry Cohen writes of the times when "a great talent, already operating at its presumed optimum, suddenly bursts beyond it and remains frozen in our memories." Tennis champion Arthur Ashe calls it "being in the zone." The great running back O. J. Simpson describes the experience of his most spectacular runs as somehow watching himself in a daydream. Recalling a sixty-four-yard open field run that gave USC a pivotal 21–20 victory over UCLA in 1967, Simpson says, "Ah, man, I just took off and ran. When it was over, I felt good. But somehow I knew it was going to happen, even though it was a spontaneous thing."

"It's hard to explain when things are going well," says playwright Neil Simon about the best moments in his work. "It seems effortless; you're very loose. Everything just seems to come easy. It's like playing your

* Jerry Cohen, "What Makes Them Winners," *San Francisco Chronicle* (Feb. 3, 1977).

best tennis; none of your muscles are tight, and it seems like you can just keep hitting balls all day."

Lyricist-director-playwright-screenwriter Alan Jay Lerner takes it a step further: "When things are going well, the elevator arrives sooner, room service is faster, you get a taxicab easier, your wife seems to love you more. You love the whole world more. I think your whole constitution changes."

These "winners," according to writer Cohen, are a species apart. "They vibrate to a drumbeat that separates them from second-best." There is truth to this notion, however vulgar and romantic. But perfect rhythm is not exclusive. According to the formulation that has emerged in these pages, it is always silently pulsing at the heart of each of us, available to our awareness on certain rare occasions. These occasions generally seem to require something out of the ordinary—an episode of danger or passion, a long period of intense concentration, a drug experience, or the total exhaustion of mental and physical resources. The sudden approach of death is often enough to do it.

For example, one afternoon not long ago, a beautiful middle-aged woman I shall call Julia was told by her doctor that a lump in her breast might well be malignant. "What flashed through my mind," she said, "was the death of two close friends who had started out the same way, with that news. At that instant, I just moved one place over, just like that. It was as if I was walking next to myself. The real me was standing next to the one who was doing things."

Julia was self-employed as an industrial consultant. The day after her doctor's appointment, she was scheduled to give a workshop with a male colleague in a neighboring city. He picked her up and they started the long drive through rush-hour traffic. Normally she dreaded those drives; her colleague tended to be morose early in the morning. On that day, however, everything was different.

"He seemed sort of soft and funny. Everything he said made me laugh. It was a beautiful April day, and all the way down I was noticing things I'd never noticed before. There was one little triangle of green lawn by an apartment building that reminded me of a fantasy I'd had as a child of a special place that would be all my own. It was the way the sun hit it, and the flowers, the kind of little white bell-like flowers that had made me dream of being a fairy when I was a child.

"Well, anyway, I went in with no notes and nothing on my mind and gave the best workshop I've ever given. Everything came out right. I couldn't make a mistake. And when they asked questions, it was as if they asked just what I wished for, just the thing to fit in with the answers I wanted to make. Somehow I knew what they were going to say before they spoke. And then I'd make the kind of answer I'd always dreamed of making—short, economical, and perfect.

"All that day, from the moment I awakened, I was surrounded by this *space* that was absolutely *clear*. Just beyond that clear space, everything was muted. But right around where I was, was clear. It's hard to explain, but it was very distinct, as if you could step into it."

By the time Julia and her colleague headed home, the day had become very hot, so they drove off the expressway for a soft drink. "We pulled right into a place that had just opened. It was made of concrete blocks, and done inside in fuchsia and pink with huge bouquets of plastic flowers on Formica tables. It was incredible in its pretention and bad taste. The waitresses were made of hard rubber. And it was perfect. It delighted me. I was in a state of total delight.

"Later, we were driving by the bay and I saw a long-necked white bird skimming along, going right with the car, just a hairsbreadth from the surface of the water. It was so perfectly mathematical. I couldn't imagine the math it would take to explain the precision of that flight. The bird was there for me to see, very personal, doing this heroic feat of flying for itself and for me. We were connected. Everything was connected."

Julia learned that she did not have cancer. The threat of death receded, and with it the sense of perfection. "But you know," she said, "sometimes I can still call back the feelings of that day for brief periods. Just that has changed my life."

The philosopher Heidegger has said that we are forgetful of existence, that we spring into full awareness only when confronted with death or poetry. But there are surrogates for death—for example, total exhaustion and the loss of some hard-won competence. The man named Richard who later became an aikidoist made it to a pre-Olympic track meet in Mexico City in 1967 as a college sprinter. The qualifying heats and the finals all came up within two days of his arrival.

"It was a tremendous strain," Richard recalls. "We didn't have time to get used to the altitude, which is over eight thousand feet. It was disorienting. Here we were in a strange city, in a strange climate, eating strange food. The day of the finals, I ran *six* races— quarterfinals and semifinals. When I walked up to the starting blocks for the finals of the hundred-meter dash, I realized I was in a state of total exhaustion. I had *nothing* left. I couldn't call on my physical resources. I couldn't call on my previous training. So I just said, 'I surrender. I give in.'"

At that point, there was a sudden shift in Richard's consciousness. "There was this strange sense of *completion*. It was as if I was in the exact *center* of time; there was just as much in front of me as in back of me. It's hard to explain. All potential in front of me was equal to all time in back of me, and I was balanced in the center. Having given in, I didn't have to relate to winning or representing my country. I was just *there*.

"The starter's gun went off and I watched myself run the race, running in the center of time. Everything was so precious—my breath, the sky, my relationships, my life. At the finish, I realized I'd won, but that wasn't the most important thing. When the judges surrounded me at the finish, I wanted to tell them about my experience. I started trying to talk in Spanish. It wasn't *winning;* it was the whole thing. I wanted them to know."

When we are pushed beyond normal limits, we don't have to be superstars or World Class athletes to touch the rhythmic core of our existence. I had one such experience while hiking on a foggy winter day. My oldest daughter, then in her early twenties, happened to be leading our group when we came to the place where the trail slants steeply downward, following the course of a mountain cataract for nearly a mile on its final plunge to a lake.

Instead of slowing her pace as the trail became steeper, she broke into a run and, as I was to write later, several members of our party followed.

> After only a few steps, it became apparent we were going too fast to stop. The way down was crooked and often twisted back on itself. There were roots, loose rocks, slippery spots. A few fallen trees lay across the trail. I was aware of these hazards, but almost immediately realized I

could not afford to *think* about them. If I was to avoid a spectacular fall, all I could do was surrender myself to the total environment—to gravity, air, fog, trees, rocks, roots, earth, the sound of water, breath, movement, the multitudinous rhythms that linked me to *all of this.* Without becoming passive in the least, but rather entering an unfamiliar mode of heightened awareness, I *let* my legs carry me down the trail at a dead run. I *let* my feet pick out the best spots. I *let* myself fly up over the fallen trees.

We came to a more level stretch of the trail and I had a chance to slow the pace. But my daughter only flew faster, and so did I, sailing down the next segment of steepness in perfect harmony with this winter rain forest, one with the white water, the liquid air, the fine jeweled droplets on the ferns and the laurels. Too soon, we came to the bottom. And there a cluster of laurels shimmered—across a deep round pool at the foot of a waterfall and also in me—cluster of silver-green leaves against black shadows, as bright as Christmas.*

I remember standing there, breathing deeply, entirely satisfied to be where I was, pleased with each moment, wishing for nothing. Like an animal or a primitive hunter after a hunt, I was without doubt or reservation or judgment. My life, for those few moments, was as clear and resonant as a crystal.

Nature offers all of us a chance to enter the crystalline state. To leave the disrhythmic city streets for some deserted wood or meadow or seashore is often

* George Leonard, *The Transformation* (New York, 1972), pp. 65–66.

enough in itself to trigger a period of perfect rhythm. An episode of this type came to my attention in a letter from a young man in Florida telling me of five days in his seventeenth year, during which he and a friend camped on a deserted island off the east coast.

"The first day," he wrote, "we set up the tent and went exploring. Toward dark, we found an oyster bed and began eating. That night, the temperature was near twenty. The next morning, looking out of my tent at the frost, I had the sense that the whole scene was extremely beautiful. I even noticed how the individual rays of sunlight seemed to pulse into the clearing. Everything was extremely clear and pleasing—even walking and breathing! This was the way we lived for the next three days; we had no arguments, no complaints, saw no people except a few boats going by in the ocean.

"The whole time I felt it was harmonious at all levels. I felt everything happened for a reason, had purpose—that everything was happening as it should, even the 'bad' things in the world. Everything was going the way it was supposed to. I experienced a deeper peace than any I had ever known. Even physical exertion was restful, in that something that normally was always irritating, driven, was at peace, interacting peacefully, harmoniously.

"Even though physically separate, I 'knew' a tree, grains of sand, sea, flying birds. Everything was God, holy; as God is total, so the driftwood branch was holy. This must be the stuff religion is made of. Never before or after have I felt so alive. I was *now,* not a step behind or thinking about the future. I was flowing as the sea. I could have blown into your eyes like sand.

"The last day on the island, we both got up before the sun and went to the salt marsh which was at low tide—about a mile wide with a snaky tidal creek running up the middle. As we stood facing eastward out to sea, the great orange-red sun lit the face of the earth, and our hearts; we rode up with it. This was our timeless moment. Had I lived on the marsh from birth to death, I couldn't have known it or myself or this earth as I did that sunrise. My first new morning, a knowing beyond words or emotion of everyday.

"On the walk back, no speaking, a light and a lightness. We walked through a flock of seagulls that a few days earlier would have noisily taken to flight at our approach.

"After hitchhiking to the car, I had trouble driving back because of my 'highness.' Once home, it was walk the dog and take out the trash. After? Nothing—person, place, drug, or sensation—has ever come close to that journey. I suppose when I settle down, I might get something accomplished, but will I ever be able to dwell with that life, God itself? Can I break out of myself again, to where I want to be?"

This young man's sense of "knowing beyond words" brings to mind the "noetic quality" proposed by William James as one of the key elements in any mystical experience. Other events and feelings presented in this chapter might also fit into the religious or quasi-religious categories listed by James in *The Varieties of Religious Experience* and by Aldous Huxley in *The Perennial Philosophy*. Episodes of perfect rhythm are obviously "religious" in the broad sense. But I want to remove them from any narrowly focused definition of that word, so that we can seek such episodes, not just in monasteries and ashrams, or even in arcadian solitude, but also in what might seem the most unlikely places—for example, at a Board of Directors meeting at a huge corporation.

Hank, a chunky, studious-looking man in his late thirties, was an executive for a major shipping firm that faced a financial crisis. He and the president of the firm conceived a daring plan to build three ships at a cost of $200 million and vie for a substantial share of the Caribbean trade. The plan would have to be presented for approval to the Board of Directors of the parent organization, a giant multinational corporation. The board meeting was scheduled in just a month, so Hank, the president, and four other executives settled in for a crash effort.

"We worked sixteen hours a day, seven days a week, for a month," Hank told me. "We had a tremendous amount of data to analyze—an overwhelming amount. We had to take the data and make a convincing presentation to the Board. We worked from eight A.M. until as late as two A.M. We'd send out for food and sit right at our desks and eat. I got too keyed up to sleep. When I'd go home to bed, I'd just doze fitfully. I didn't relax until it was all over, then I collapsed at my desk one day. Our team got very close during that period, and we still talk about the prodigious amount of work we did."

After this period of "tension, personal clashes, terrible disappointment, and elation," Hank and the president flew to New York with the completed proposal. Hank had the job of making the presentation.

"It was like my whole life was wrapped up in that one moment. When I stood up, it was like in the movie, *Patton,* when the general was standing on a battlefield and felt that all through his life—and maybe through past lives—he was meant to be right there, on that spot, at that time. It was the same with me. I had a feeling it was predestined, as Patton did.

"I recall the Board of Directors—thirty people seated around this tremendous table. The vice-chairman was reputed to be a financial wizard, with great respect from Wall Street. But I knew I had everything in the palm of my hand. I always knew that some-time in my life there would be a tremendous struggle and that I'd come out on top, with a feeling that I couldn't possibly fail. And now I was in the right place at the right time. Predestined.

"The presentation lasted an hour and a half, and everything just fell into place, like A plus B plus C equals D. Like a mathematical formula. Sometimes I got the feeling I was lecturing college students on a perfect corporate plan. They were listening that re-spectfully. There was no opposition. I felt no matter what they asked me I could have been right on top of the situation. If they had asked me the question in Swahili, I felt I could have come up with the answer.

"On the plane after the meeting, there was a strange sense that everything was *right*, everything that was and could be."

For most of us, such episodes are rare. For Hank, it was once in a lifetime. The plan turned out well; he was promoted to vice-president. But these outcomes later seemed almost incidental to the month-long ordeal, the predestined moment before the Board. Perfect rhythm. Now Hank has "success," a weight problem, and one shimmering memory.

The moments pass. We are left with a sense of life's riches, a sense of loss, and a tantalizing question: Why can't it always be that way? Thomas Traherne, the seventeenth-century mystic poet, described the riches and the loss:

The dust and the stones of the street were as precious as gold. The gates at first were the end of the world. The green trees, when I saw them first through one of the gates, transported and ravished me; their sweetness and unusual beauty made my heart to leap, and almost mad with ecstasy, they were such strange and wonderful things. The Men! O what venerable and revered creatures did the aged seem! Immortal Cherubim! And young men glittering and sparkling angels, and maids strange seraphic pieces of life and beauty! Boys and girls tumbling in the street, and playing, were moving jewels. I knew not that they were born or should die. But all things abided eternally as they were in their proper places. . . . And so it was that with much ado I was corrupted and made to learn the dirty devices of the world. Which now I unlearn, and become as it were a little child again, that I may enter into the Kingdom of God.

During the periods of perfect rhythm, what is most ordinary becomes most wondrous. Nothing exotic or fantastic is needed. There is, in Blake's words, "a World in a Grain of Sand,/And a Heaven in a Wild Flower."

Most of all, the experience stands entirely beyond judgment. A judgmental attitude is, in fact, proof against perfect rhythm. To judge experience as it is happening leads you to judge experience before it happens—premature judgment, prejudice. In this regard, Archie Bunker and a judgmental college professor differ only in the quality of their judgments; they share the same essential modus operandi. To approach events wide open and without judgment is to achieve the first condition leading to perfect rhythm.

A man I shall call Ezra is a professor of comparative literature and a sometime writer of articles for the quality magazines. On assignment for one of these magazines, he subjected himself to an intense encounter-group experience at Esalen Institute on the California Big Sur coast. Despite his attempts to maintain a certain ironic distance, he found himself being swept into the group by the power of his own unexpressed emotions.

On the third day, he recalls, there was a great surge of sadness. "It was something like grief and fear of the grief. On the next afternoon, it was as if all that was negative and repressed in my life achieved a critical mass—grief, anxiety, terror. Then there was an explosion. When the critical mass burst, I felt life suffusing me. My vision was absolutely transformed. I began to see things as if for the first time, like someone who suddenly has his sight restored. I lost all fear, all anxiety, all concern about the future. I had no grievances. I felt loving. I felt wide open. I had almost immediate access to my feelings. Everything I uttered seemed absolutely honest and true. There was no one I shied away from, no experience I avoided.

"I began looking at everything around me—the trees, the path, the pebbles on the path. I realized what it was like to be in the here-and-now. It was the same with people. No one could possibly bore me. Every voice was a thrill. Every moment of being alive was a joy. Nothing could possibly go wrong.

"That night I must have sounded like a lunatic to someone who had just come in. I was hugging everyone—a sappy nut. I would have been disgusted at myself in retrospect. I felt I had to give everyone the message. Those last lines in Kubla Khan—

> *And all should cry, Beware! Beware!*
> *His flashing eyes, his floating hair!*
> *Weave a circle round him thrice,*
> *And close your eyes with holy dread,*
> *For he on honey-dew has fed,*
> *And drunk the milk of Paradise.*

—I was like that—a crazy man, a lunatic. I would have run from that creature. But, God, I'd love to have that feeling forever and ever.

"Just a week earlier, I had been full of plots to avoid people and situations, and to preserve my integrity. The metaphor I'd use for the change is the oldest one: being asleep and being awake." Ezra smiled with faint irony. "Yes, I was reborn—a reborn professor."

The feeling remained with Ezra for about two weeks and then began petering out. "I realized that in order to keep that feeling I'd have had to change my life. I'd probably have had to get a different job. I'd have had to live differently in my house. And then, too, I started to write about it. I knew I was taking a risk of losing it; writing is essentially not a here-and-now thing. So I gave the feeling up. In order to maintain myself as I had been, I've had to sacrifice what might be."

Ezra paused for a moment, then continued thoughtfully. "It's really about *time,* isn't it? Time and rhythm and the flow of time. When I first arrived at Esalen, I had an unpleasant exchange with a large, threatening-looking man. We both had rooms in the same area, and I'd frequently meet him on the trail. I'd see him coming on the other side of the canyon, and I'd start rehearsing scenarios and I'd suffer a thousand deaths until we passed.

"Then, after my experience, I saw him in the distance, and I thought, *I'm not there yet.* I realized I'd been missing everything on the way worrying about him. So this time I paid attention to the pebbles, the undulations of the trail, the place where it was stony, the rivulets of water, the butterfly tree.

"When I got to him, it was a surprise. I said, 'Hi,' and passed. I realized he was a very warm and likeable person. I realized that previously I had been *dead* right

129

up to the moment we met. I had been *not* in my body, *not* on the path.

"You see, it was two different kinds of time. Before my experience, I'd see the space between us and I'd figure how long it would take before we'd meet. It would become linear, mechanical. *I'd* become mechanical. After my experience, time changed. Every moment was full, and every moment had its own time. If it took, say, one second to experience a particular pattern in the trail beneath my feet, the experience in terms of its intensity was not commensurable, not measurable in seconds. Time for me was no longer a measure of length, space. The measure of time was intensity. The moment in which I looked at him was perhaps less intense than the moment I looked at the butterfly tree. So time, as we know it, was not a primary quality but a secondary quality, like color or sound."

Ezra paused and shook his head. "I lived that way for two weeks, without fear, without anticipation, with my eyes wide open, flowing with time. But I couldn't leave the security of my bondage, so I lost it.

"The profound wisdom I wish I could accept as the ground of my own action is simply to accept what is and not approach events with judgment—not giving reasons why, not stating a case. But to do that, I'd have to walk on a different path, not knowing where it ended.

"There are indications the situation is impossible to maintain. Maybe the best we can hope for is to have it now and then."

For most of us, as we have seen, the episodes of perfect rhythm are hard to come by, harder to hold. Though there are guidelines, there are no teachers, no books, no instructions that can absolutely guarantee those times when every moment is a joy and nothing can possibly go wrong. Perhaps it is enough to know that such episodes exist, and that they can occur when we least expect and most need them.

So we could leave it at that. We could let the notion of perfect rhythm stand like a beacon on a hill, a distant, elusive measure of whatever efforts we might make to find the natural beat of our own lives. But there is that other possibility: Perhaps, after all, perfect rhythm is always present in our every action and relationship, and it is only our awareness of it that is a microsecond out of phase. Could it be that we miss the experience not because it is so distant but because it is so close? Let us now examine our own unacknowledged powers, and discover what this elusive perfect rhythm is all about.

NINE

Intentionality and Power

On certain fine moments, most of us feel we are "in sync" with everything in the universe. These moments, as we've seen in the preceding chapters, are quite real in terms of experience. The world and our relation to it seems remarkably different; our connectedness with all we perceive is vivid and undeniable. I've theorized that this synchrony, this silent pulse of perfect rhythm, is always present in each of us; that it is, in fact, the most essential and irreducible condition of our existence. Awareness of perfect rhythm generally occurs when, for one reason or another, we are stripped of the barriers of ego, custom, language, and judgment, and find ourselves at the dimensionless point from which space-time unfolds. At this point, which we term "the present moment," it is as if we were involved with the process of creation, instant by instant, of all that is.

But is the experience of perfect rhythm something that "just happens," or can it be intentional? Is it purely subjective, or can it result in objective, measurable changes? And are there ways of getting in touch with and using the silent pulse of perfect rhythm?

Throughout the ages, saints and yogis, through strenuous discipline, have sought to achieve what I call perfect rhythm, and it is becoming increasingly difficult to dismiss feats of yogic control, faith healing, fire walking, and the like. But here, as in the last chapter, I want to turn to something far more commonplace: a situation in which we are tricked into using the extraordinary force of our intentionality.

This situation involves a truly wondrous drug—an embarrassment to pharmacological researchers, a godsend to physicians since the time of Hippocrates. The name of this drug is *placebo,* and it is the "placebo effect" that proves beyond all doubt the powers of what we carelessly call "mind" over matter. The word comes from a Latin verb meaning "to please," and some of us might think of the placebo simply as a sugar pill, a fake medicine given by a lazy doctor for an imaginary ailment. It's true that the placebo is most commonly a milk-and-sugar tablet disguised to look like an authentic pill, and some doctors do prescribe it simply for lack of anything better to do. But the placebo itself is merely a formality, a means for sealing a pact. The sugar pill might as well be a saline injection, a real aspirin, a surgical procedure, or, in other cultures, a shaman's chant or a potion of lizard's blood. The placebo effect derives not from the potion but from the process, which is one of authorization. The healer simply authorizes the patient to do what he or she is already easily capable of: that is, to control even the most esoteric bodily functions, to grow or destroy tissue, to produce sickness or health.

Each culture has its own expectations regarding the

healing process. Ours involves a visit to the physician, a sympathetic and informed audience for our complaints, some sort of examination, and, at the least, a written prescription. Taking the prescribed remedy can set the healing process into motion. If this process, in the case of a placebo, involved merely a strengthening of the will to live or a generalized "getting better," it would deserve little of our attention. But there's far more to it than that. The placebo effect, as recent experiments are beginning to demonstrate quite clearly, has amazing power and specificity.

Placebos have been shown, for example, to be 77 percent as effective as morphine in relieving postoperative pain; the more severe the pain, the more effective the placebo. A test with eighty-eight arthritic patients revealed that a placebo is equally effective with aspirin and cortisone in the relief of symptoms. What's more, the benefits from the placebo in these cases included improvement in sleeping, eating, elimination, and reduction of swelling. Heroin addicts, given saline injections as placebos, continue to show the same response as with the real drug; withdrawal effects don't start until the placebo is stopped. Blood-cell count, respiratory rate, vasomotor function, and hypertension are among the functions that can be strongly affected by the administration of placebo.

The placebo effect, as previously suggested, can be extremely precise. The patient who is told what to expect from a certain drug can produce that exact effect when given a sugar pill, down to creating a surplus of specialized blood cells called eosinophils. Just as a placebo can produce healing conditions in the body, it can also produce the side effects of the drug it mimics. Of one group of persons given a placebo in place of an antihistimine, 77.4 percent reported drowsiness, an antihistimine side effect. *Saturday Review* editor Norman Cousins reports on a study of the drug mephenesin's effect on anxiety:

> In some patients, it produces such adverse reactions as nausea, dizziness, and palpitation. When a placebo was substituted for mephenesin, it produced identical reactions in an identical percentage of doses. One of the patients, after taking the placebo, developed a skin rash that disappeared immediately after placebo administration was stopped. Another collapsed in anaphylactic shock when she took the placebo, and again when she took the drug. A third experienced abdominal pain and a buildup of fluid in her lips within ten minutes after taking the placebo—before she had even taken the drug.

You might wonder why we can't produce these remarkable results without being deceived by physicians or medical researchers. The answer is that we can. It's just that our current medical model is based on external authority rather than individual self-responsibility. Cousins tells of a study in which patients with bleeding ulcers were divided into two groups:

> Members of the first group were informed by the doctor that a new drug had just been developed that would undoubtedly produce relief. The second group was told by nurses that a new experimental drug would be administered, but that very little was known about its effects. Seventy percent of the people in the first group received significant relief from the ulcers. Only 25 percent of the patients in the

second group experienced similar benefits.
Both groups had been given the identical
"drug"—a placebo.*

Sometimes hypnosis enhances the placebo effect,
but concentration and conviction, with or without
hypnotic induction, is the key, as was shown in an
experiment with thirteen Japanese high school students who were highly sensitive to a poisonous plant
similar to our poison oak. Five of the thirteen were
hypnotized by a physician, then, with eyes closed, were
touched by leaves from a harmless tree while told they
were being touched by poisonous leaves. The other
eight students were not hypnotized, but simply told,
falsely, that the leaves they were being touched with
were poisonous. All thirteen developed some degree
of the familiar poison oak-type inflammation. The
same students were touched on the other arm by the
poisonous leaves while being told the leaves were
from a harmless plant. Four of the five hypnotized
students and seven of the eight who were not hypnotized did not develop any inflammation on that arm.†
  The implications of such experiments are so pro-

* See Cousins' article "The Mysterious Placebo: How Mind Helps
Medicine Work," *Saturday Review* (Oct. 1, 1977), from which the
preceding examples were taken. Cousins himself was involved in a remarkable self-cure of a rare, debilitating disease, ankylosing spondylitis,
from which, according to one specialist, he had one chance out of 500
for recovery. With the cooperation of his own physician, he moved
from a hospital room to a hotel room, dispensed with all drugs, prescribing for himself instead massive doses of vitamin C and laughter.
To make himself laugh, he read books of humor and projected films of
old *Candid Camera* shows in his room. Recovery from the supposedly
progressive and incurable disease was slow but steady. He now lives a
normal life.

  † Y. Ikemi and S. Nakagawa, "A Psychosomatic Study of Contagious
Dermatitis," *Kyushu Journal of Medical Science* 13 (1962), pp. 335–50.

found that it's hard for us to take the next step in understanding. This step requires at the least that we set aside the conventional separation between "mind" and "body" and that we consider instead the possibility of an *intentionality* that operates across the entire realm that we identify as body, mind, and spirit. Intentionality can be thought of as the vector of identity. It always works initially in terms of *structure* rather than of the particular material through which the structure manifests itself. A radio receiver is a structure; so is a wiring diagram for that radio, and so is that diagram held in the memory. It's far faster and easier to make changes in this structure by shifting a thought or making a mark on a piece of paper than by manipulating the radio itself. In all three cases, however, intentionality is primary and irreducible. The body is also a structure, infinitely more complex than any radio, and the remarkable changes we've seen here must go back to the *intention to change*. This intention, as demonstrated by the placebo effect, is not necessarily lodged in conscious thought or will.

Conscious will, in fact, is only one of the instruments of intentionality. To the extent that it considers and judges, consciousness is indeed rather ineffective for these purposes. Effective intentionality springs not from ego but from identity, which transcends categories such as "body" or "mind." It exists in the here-and-now, in what Alfred North Whitehead calls "the individual immediacy of an occasion." Intentionality can be held in the vessels of consciousness, memory, and the will, but it creates transformations only in the moment, in the domain of the silent pulse, which ordinarily escapes our attention. When we can bring our apprehensions down to this place, down to the point of interplay between identity and holonomy, then there is a universe for us to experience. It is here, too, that intentionality works in remarkable and sometimes mysterious ways.

On March 5, 1962, just twenty-one days before Easter, a ten-year-old black girl began bleeding from the palm of her left hand while in her classroom in Oakland, California. Over the next nineteen days, she bled up to five times every day from wounds in both feet, both hands, the thorax, and the middle of the forehead. The bleeding was witnessed by her parents, classmates, teachers, and the school nurse, as well as by the physician who reported the phenomenon and other hospital staff.* On Good Friday (the nineteenth day of bleeding), the girl bled simultaneously from all six wounds. When it stopped that day, she felt "it was all over," and indeed no further bleeding occurred.

The girl comes from a highly religious Baptist family, but she had had no prior knowledge of the Christian stigmatic phenomenon. About a week prior to the bleeding, she had read *Crossroads* by John Webster, a book about the crucifixion. Four days before the bleeding began, she had watched a television program on the crucifixion that had produced strong emotions and a vivid dream that night. Auditory hallucinations that began a few days before the onset of the bleeding continued until Easter Sunday, and generally consisted of simple, positive statements, such as, "Your prayers will be answered."

And so this ten-year-old stigmatic from Oakland, California, joined countless thousands of others, in and out of formal religious institutions, who have shared in the suffering of Christ. The phenomenon, as in her case, generally follows intense contemplation of that suffering and manifests itself in strikingly literal

* Loretta Early and Joseph Lifschutz, "A Case of Stigmata," *Archives of General Psychiatry* 3 (February 1974), pp. 197–200.

reproductions not only of the wounds inflicted by nails and lance but, in some cases, the bruise on the shoulder incurred from carrying the cross, the chafing of bound wrists and ankles, and lacerations around the head caused by the crown of thorns. Nuns and other highly religious women have also produced the "token of espousal," a spontaneous modification of the flesh to resemble a ring around the finger of one who has experienced ecstatic betrothal with Christ.

There is no disputing the power and specificity of these phenomena, only their origin. A religious fundamentalist might argue that the wounds come literally from God, a gift to the worthy. A person on another side of the belief chasm might resort to quasi-scientific buzz-words in the way of explanation: "suggestion; conversion hysteria." These extremes are reconciled in the holonomy-identity formulation.

Look at the girl from Oakland as a holoid of the universe experienced from a unique identity or point of view, and as a context rather than a content. A context is not a container. The word comes from the Latin terms *con* and *texere,* meaning "to weave together." Context, then, is a process of relating, of weaving together. This context that is an individual human being manifests itself as body, mind, soul, history, works, and the like. (Each of these manifestations, as we have seen, can ultimately be treated as wave phenomena.) Consciousness, which is tied to language, convention, and ego, has access to only a minuscule fraction of all the material available in the holoid.

The story of the crucifixion exists in the universe; therefore, it exists holonomically in the girl, whether consciously known or not. In this case, consciousness of the crucifixion (but not of the stigmatic phenomenon!) triggered the process of focused concentration.

To achieve the results seen here, however, the process
that began consciously must go deeper than language,
convention, and the ego until it reaches the wordless
interplay between holonomy and identity, where reality is continually creating itself, bursting into existence
in terms of objects and events that may or may not rise
to conscious awareness. This is the point of relationship at which intentionality operates, whether consciously or unconsciously, the point where "all things
are possible" within the fundamental constraints of
the universe.

We might bear in mind that intentionality operates
not primarily in the realm of objects and events, but
in the realm of process, in the moment. Objects and
events are then perceived to have changed to some
extent. The *results* of intentionality are thus always
history, if history only a billionth of a second long. The
crucifixion—to repeat a key point of this formulation
—exists in all its fullness in the context which is the
girl from Oakland (and also in the context which is
each of us). Through a process of intensely focused
contemplation, the girl experiences the crucifixion in
the silently pulsing present moment. The history of
that experience appears in the body-as-object. Questions arise: Did God cause the wounds? If you wish,
yes; God, Christ, the crucifixion—all were fully represented in the experience. Did the girl cause the
wounds? Yes, this is also true; the intentionality involved was seemingly most closely associated with her.

To theorize about "cause," however, leads only to
endless thickets. Philosophers since Aristotle have argued about causation with no satisfactory resolution.
Most now agree that "cause" as used in public health
("Poor sanitation measures *caused* the cholera epi-
demic"), medicine, law, military planning, and the like
can be useful in assigning responsibility and taking

action within those spheres, but that the same concept blurs and misleads in more fundamental matters. In this instance, I want to set aside the question of causation and simply say that intentionality, whatever its source, was closely associated with and necessary for the occasion from which the wounds arose.

In this view, to go a step further, the particular mechanism responsible for the appearance of the wounds is of secondary importance. Scientists might devote themselves, appropriately, to discovering by what means the body produces a stigmatic wound—whether, for example, by an alteration in local blood flow or by some sort of breakdown in subcutaneous tissue. This discovery would be interesting and significant. But whatever the mechanism, even if the girl produced the wounds on herself with a sharp instrument, we can still view intentionality as being primary in the alteration of bodily structure.

POSITIVE TRANSFORMATIONS

Conscious or not, intentionality is continually in the process of influencing the universe that is each of us. A sobering thought! For if intentionality can produce bleeding wounds, can't it also produce more serious, even fatal, breakdowns in our bodily systems? Yes, this is surely happening all the time. Few, if any, medical scientists today would deny the powerful, if not decisive, effect of "attitude toward life," "state of mind," or "life-style" in producing pathology. Almost every illness—from hemorrhoids to heart disease, from the common cold to cancer—is attributed by some physicians to psychosomatic factors.

How about the other side of the coin? It's obvious that intentionality can damage the body. Can it also repair damage? Can it produce positive bodily transformations, exceptional states of good health? If the

141

positive possibility seems far more remote than the negative, it's perhaps because medical science has customarily turned its attention and energy—its intentionality—toward pathology. Extraordinary good health, much less positive bodily transformation, stands for the most part as a terra incognita. Little thought and few experiments have gone into this area, which has become somehow suspect. It's as if the "power of mind" to sicken and destroy is a respectable concept, easy for us to deal with, while the same power, applied to healing and transformation, smacks of charlatanism and the occult.

For the hard-nosed medical experimenter, then, the placebo effect in its positive sense is a nuisance to be minimized or avoided. This is accomplished by the "double-blind" method of evaluating a new remedy: Neither the experimenter nor the patient knows whether a particular dose contains the medicine or a placebo instead. The patients' responses to the course of treatment are carefully noted, and only after the experiment is completed are the responses to medication and placebo compared. Any consistent difference is then attributed to the pharmacological action of the drug. This action, with or without the placebo effect, can indeed produce certain bodily changes, and to isolate it is useful. But "double-blind" has taken on such a magical quality that it has created a virtual taboo among many medical researchers on experimentation concerning the powers of intentionality. Significant experimentation remains to be done on ways to enhance intentionality, not only in the absence of active drugs, but also in conjunction with them.

The truth of the matter is that the placebo effect can never be totally excluded. It contributes to the effectiveness of every drug, every medical procedure. The main problem with the placebo itself, whether used in a double-blind experiment or prescribed by a doctor as a therapeutic agent, is that it involves deception. The doctor writes a prescription for a sugar pill and thus authorizes the patient to set his or her transformative powers to work. In doing so, however, the doctor damages the field of relationship between the two of them, and thus reduces the possible effectiveness of the process. This damage carries over to the next interaction and the next, shaping a dominance-submission, parent-child relationship that has its own obvious limitations.

The placebo effect works best when both the patient and the healer are convinced of the power of the treatment. This is true in the case of a medicine man offering a potion of lizard's blood. It is also true of a medical doctor prescribing a tonsillectomy. Science can prove that lizard's blood is medically ineffective and perhaps harmful. The same thing is now being proven about most tonsillectomies. Fashions in therapeutic measures change over the years. Respected physicians were once convinced of the value of bleeding their patients; in the year 1827 alone, France imported 33 million leeches after its domestic supply was depleted. Our modern hospital, with its heavy emphasis on surgery, technology, and rigid scheduling, may seem as irrational to future generations as leeches are to this one. Both thrive on belief.

The next step in reowning our powers might involve doing away with the pill or other physical remedy, or simply treating it as ceremonial (which is not to say without real powers; just the opposite). After that, we might come to realize that the presence of a second person, a doctor or healer, while useful in shaping the occasion, is not absolutely necessary. Every one of us is continually influencing our universe through our own intentionality. We've become accustomed to thinking that we must go to a doctor and get a prescription before putting our transformative powers to work in that aspect of reality that we call our own bodies. We've also been involved in an unspoken social agreement to use these powers, once released, only toward correcting pathology rather than toward developing extraordinary abilities and states of exceptionally good health. The body stands here as metaphor for the rest of our universe. When we're willing to take personal responsibility for positive transformations in this aspect of our being, we become increasingly empowered in other aspects.

This is no suggestion that we refuse to see a doctor when sick; our medical science is quite effective in cases of trauma or acute illness. It is rather a suggestion that we acknowledge our transformative powers and begin to turn them toward the positive. We have no way of knowing how far we might go in that direction simply because we really haven't tried.

An example of what might be called positive physical change is to be found in a number of experiments in which women significantly enlarged the size of their breasts through suggestion. In one carefully controlled experiment, thirteen women participated in twelve weekly sessions during which they received hypnotic induction and repeated suggestions to feel the sensations in their breasts that occurred during puberty. In the course of the experiment, the average bust measurement increased 2.1 inches, from an average of 33.6 inches (after exhalation) to 35.7 inches (also after exhalation). The rib cage just below the breasts did not show a significant change during the twelve weeks of the experiment.*

* J. E. Williams, "Stimulation of Breast Growth by Hypnosis," *Journal of Sex Research* 10 (1974), pp. 316–24.

Further examples of positive, if seemingly bizarre, physical change are to be found in the feats of Indian yogis—just the kinds of things we "sophisticated" Westerners have tended to discount or simply ignore. But there is another arena of positive physical transformation, so obvious that we fail to credit it as such, which proves the power of intentionality beyond all doubt. That field is sports, and the evidence is contained in the record books over the last century. Now that the mile is run in less than 3:50 and weight lifters can clean and jerk more than 560 pounds, these feats are not called supernatural. But if you had told a sports expert of the year 1878 that such performances were humanly possible, he would have thought you quite mad. In recent years, as a matter of fact, a fifty-year-old man has run the marathon faster than the 1948 Olympic gold medalist in this event and a sixty-five-year-old man has bested the time of the 1908 Olympic marathon champion.

Some of this truly fantastic improvement can be attributed to "technology"—better selection, training methods, nutrition, vitamins. But the same kind of technology has been applied to racehorses, plus breeding practices that would be out of the question with humans, with no such improvement in performance. Today, athletes in every field from golf to weight lifting are unequivocal in stating that "mental" or "psychological" factors are primary in achieving top performance. Technical measures are involved in building the base camp. But the journey from there to the peak is all intentionality.

"The mind is the limit," says weight lifter Arnold Schwarzenegger, five times Mr. Universe and perhaps the world's premier body builder. "As long the mind can envision the fact that you can do something, you can do it—as long as you really

145

believe 100 percent. It's all mind over matter."

Schwarzenegger talks of *thinking* his way into a certain muscle and visualizing it as larger, and of lifting weights mentally before physically. "When weight lifters are standing in front of the bar," he says, "they must, in their minds, lift it in order to then lift it physically. If they have one percent doubt, they can't do it."

This sort of visualization is now being practiced in many sports, especially in East Germany and the Soviet Union. It's not simply a question of "giving it the old college try." "Trying" in the ordinary sense can even be detrimental. The practice, rather, entails creating a sense of the event that is vivid and fully realized, an occasion in itself. It's not even necessary to *see* the accomplishment as a picture in the mind's eye; we speak of visualization simply because the visual system is prominent in the human animal and thus easy to talk about. (Champion golfer Jack Nicklaus never hits a shot without "seeing" the ball exactly where he wants it to finish, "nice and white and sitting up high on the bright green grass.")*

Whatever visual or feeling language we use, the key lies in making the occasion as real and present in the realm of intentionality and structure as in the realm of energy, matter, space, and time. "All I know," says Schwarzenegger, "is that the first step is to create the vision, because when you see the vision there—the beautiful vision—that creates the 'want power.' For example, my wanting to be Mr. Universe came about because I saw myself so clearly, being up there on the stage and winning."†

* For numerous examples of intentionality at work in the sports field, see Michael Murphy and Rhea White, *The Psychic Side of Sports* (Boston, 1978).

† Schwarzenegger quotes from an interview, "The Powers of Mind," by Ken Dychtwald, in *New Age* (January 1978).

I learned aikido from a teacher who operates from the premise that the perfect move, the perfect throw, *already exists.* Our mission was simply to join it. Before being able to do so, we would have to practice long and hard, master the basics, take our share of bumps and bruises. But through all of this, the perfect throw would be there—in the process, the eternal present— always there for us to join. Sometimes I have my own students hold the vision or the feeling of a certain throw in their minds, and then practice it over and over again for an hour, until they are drenched with sweat and barely able to move. Too tired to *try* any longer, some of them undergo a kind of transformation. It's as if you could see their hesitancy and self-consciousness peeling off like old, dead skin, and something radiant and newly energized emerging. For a while, the extraordinary becomes commonplace and relatively inexperienced students perform like masters, realizing the grace, the perfect rhythm that has been theirs, unclaimed, through all of eternity.

FOCUSED SURRENDER

Again and again we encounter this paradox: intense effort that becomes effective only through total surrender, the unlikely marriage of trying with not-trying, during which intentionality can alter structure. From 1973 through 1975, for instance, a researcher named Duane Elgin conducted a remarkable series of exercises at Stanford Research Institute, attempting through intentionality alone to influence a sensitive, heavily shielded magnetometer. This instrument measures changes in magnetic field, and records these changes on a moving sheet of paper. The first few exercises generally followed the same course. Elgin would sit or stand a few feet from the magnetometer, where he could see the recording device, and would focus all the force of his will on the instrument, trying to influence it. He would continue this concentrated effort for twenty to thirty minutes, watching the needle tracing an almost straight line: no results. Finally, exhausted and exasperated, he would say to himself, "I give up." At the moment of his surrender, the needle would start indicating a change in magnetic field. These changes were by no means insignificant. In some of the exercises, the needle went entirely off the scale; to get such results by normal means would require a force estimated to be 1,000 times stronger than that of the earth's magnetic field. Nor did physical distance lessen Elgin's effectiveness. In one instance, he was able to affect the magnetometer strongly from his home, several miles away.

Later, Elgin learned to refine his technique. "I'd spend twenty to thirty minutes doing the best I could to establish a sense of rapport and connectedness with the instrument, and with great will and concentration I would coalesce that sense of connectedness into a field of palpable energy. I'd feel myself coming *into* the magnetic field and pulsing it to respond. Then, when there would be a moment of total surrender, the response would occur."

Here we can see a pattern begin to emerge. Elgin's state of being during his successful operations with the magnetometer—we can call it *focused surrender*—bears close resemblance to states we have encountered earlier in this book. Richard's times of grace on the aikido mat and on the track in Mexico City were characterized by focused surrender, as were the experiences of Julia and of Ezra, of Mary and Bill, and Hank the executive, triumphant in a New York board room. I can think of no better term to describe my state while running down Cataract Trail, and this paradoxical marriage between zeroing in and letting go is surely involved in the remote-viewing experiments.

In all of these experiences, the participants have had a strong sense of connectedness with objects and events that otherwise would have seemed quite separate. This makes good sense in terms of the holonomy-identity model, in which each human being *is* the cosmos from a particular point of view. Both focused intentionality and surrender of ego are necessary for experiencing existence at this fundamental level. In some of the episodes of focused surrender, the states of the participants seemed to have had noticeable effects on the world around them in ways that defy what we consider normal causality. We have no idea how far these effects, which are not forbidden by modern physical theory, can spread. Focused surrender is also involved in the placebo effect, in which intentionality has a powerful influence on the world of matter. Occurring on the level of unconsciousness, however, this focused surrender generally does not involve a conscious sense of connectedness with the cosmos.

These moments, these episodes, are not considered a part of our normal existence. Before we can realize the awesome powers of our intentionality, most of us have to be tricked by a placebo or in some way trick ourselves or be stripped of our egos or relinquish them in moments of exhaustion or isolation or passion or the apprehension of death. Why must this be? Why can't we realize our powers all the time? One reason is obvious. A stable society requires the suppression or voluntary renunciation of unlimited individual power or, at the least, an implicit consensus about just what powers can be used. Within almost all of us there is undoubtedly a strong intentionality to maintain a viable, reasonably stable society with individual freedom and privacy, not only as a support system for our own individual lives on earth, but also as a matrix for further evolution.

We've already noted the military and intelligence interest in remote viewing. Now there are persistent rumors that both the United States and the Soviet Union are experimenting with psychotronic devices, which combine "mind power" with electronics. In one scenario of long-distance hexing, radio waves are being used in some way to carry the intentions of Russian psychics to confuse and disorient Western political leaders. I tend to discount such seeming paranoia, but with some caution. It's hard to ignore the fact that the wildest paranoid fantasies of the sixties concerning unlikely and truly fiendish covert activities of our own government have turned out to be rather conservative.

What kind of world would we have if numerous people gained the power to influence instruments or minds at a distance, or bend metal without physical force? The dangers came sharply home to me during a recent airline flight. Shortly after takeoff from San Diego, the three-engine jet lurched alarmingly. A few moments later the captain's voice came on the cabin speakers. "I'm afraid I've got some bad news, folks," he said in a sober voice, unintentionally parodying the old airline-captain jokes. He went on to explain that the left-wing flap was stuck in the down position and that we would have to go back and land. Seated behind the wing, I could see the worm gear that extended the flap. It was long and relatively slim; a bend of only a few degrees might easily jam this mechanism.

The metal-bending demonstrations of psychic Uri Geller and others came immediately to mind. Many of these demonstrations have been revealed as mere magicians' tricks. It's extremely difficult, however, to discount the phenomenon entirely. Rigidly controlled experiments using strain gauges have shown that certain people can produce deformations in metal that can't be attributed to known physical forces.* Reports on metal bending emphasize its unreliability and indeed its capriciousness; attempts to bend one spoon reportedly result in spoons elsewhere bending. As we circled to land, I considered what might happen if the ability to bend metal by intentionality alone were developed in malicious, irresponsible, or simply uncentered people. As long as we have to depend on' metal for flight and for so much else in our complex and rather fragile technology, I thought, we'd best forego this power.

The same thing can be said about all meddling with what is "normal" and with the introduction of all new technology, whether psychic or physical. We can trace pollution, the decay of the inner city, the breakdown of the family, and the increase of youth crime and delinquency largely to the introduction of the private automobile in Western culture. How much more dramatically the development of psychokinetic antigravity, with its promise of unlimited, heat-free energy, would rip apart the fabric of all human societies! The legends of Prometheus, Faust, and Frankenstein sound the same dire warning. In her 1818 novel about the overly ambitious Dr. Frankenstein, Mary Shelley made the moral plain: "Learn from me . . . how dangerous is the acquirement of knowledge, and how much happier that man is who believes his native town to be the world, than he who aspires to become greater than his nature will allow."

It might turn out that a person's attempt to use a

---

* Experiments by Scott Hill and others at the University of Copenhagen with the French psychic Jean-Pierre Girard are particularly impressive. They suggest that stretching as well as bending is involved in this kind of metal deformation.

power such as metal bending for selfish purposes would limit that person's ability to achieve the power. We've seen that bringing intentionality into play in any deep and novel manner requires relinquishing ego. To relinquish ego is to relinquish all desire to gain power over others. To cling to ego and its selfish desires, then, might be to lose the extraordinary power of intentionality, or at least to lose it in the long run. What I have called perfect rhythm thrives in a Taoist, not a Hobbesian world. There is, of course, another kind of power going back to the French root of the word, which means "to be able." That power, the *ableness* to fulfill your bestowed mission in life, to achieve your own potential, would seem to enhance the development of extraordinary abilities. At least, we would hope that to be the case.

It's still possible to argue that "mind" will never be able to move or bend metal. But it's impossible to deny the extraordinary powers of intentionality to influence the flesh and bones and sinews and fluids of the human body, which (and we tend to ignore this fact) is also matter. In this chapter, we've noted the curious resistance and even aversion to any positive transformation of the body, any change beyond "normal." Part of the reason for this resistance lies in conventional medicine's single-minded focus on sickness at the expense of wellness. But, to be fair, there's more to it than that. Might not positive transformation of the body create monsters? There are some people who already see the monstrous in the strange feats of Indian swamis or the rippling muscles of Arnold Schwarzenegger or the prodigious heart and lungs of marathoner Frank Shorter, who can run over twenty-six miles at an incredible five-minute-a-mile pace. And these developments are undoubtedly only a beginning. What wonders, what glorious monstrosities,

await us in the future, as more and more people realize
the powers of intentionality? The conservative impulse in each of us warns of experimentation, and
counsels that we settle for the normal.

The only problem is, what is "normal?" As we look
around at the sedentary, rigid, fat-encased, nicotinic,
unbalanced, unfeeling, unfulfilled bodies that are
regularly certified by our current medical science as
"well" and "normal," we might begin to realize that
we have set our sights shockingly low, and that, in fact,
each of us is being cheated of our true potential every
day, every hour, every minute of our lives. I've chosen
the human body as the chief focus of these discussions
of intentionality because it is so complex yet unified
and coherent, so accessible, so fast and accurate in
providing feedback for our actions. But what is being
said of the body is equally true of the body politic. In
social matters, too, we generally accept conditions
that demean human dignity, much less potential; we
tolerate massive injustice, hunger, brutality, as "natural" or "inevitable." And in our interpersonal affairs,
as well, we commonly settle for stereotypes, avoid-
ances, the stale debris of our first fine hopes. Selfish
or careless use of the power of intentionality is irresponsible. But surely it is equally irresponsible to let
this world languish in its present state. Let us now look
at one of the universe's most fundamental, if only
recently enunciated, characteristics. This characteristic, when fully understood, might change our ideas
about the possibility of personal and social transformation.

TEN
The Intention
of the Universe

The glory as well as the monstrousness of the West lies in its strong sense of history, its feeling of time as alive and crackling with tensions and possibilities. For the Hindu, time is a power of deterioration in a world of illusion. Salvation means being saved from history and time, being released from the wheel of birth and death. The Old Testament prophets came up with the radically different idea of God as revealing himself *in* history—more than that, *through* history—and thus the idea of creating some kind of heaven here on earth. This idea has led to the Crusades and to Chartres Cathedral, to the Communist Revolution and to human footprints on the moon; and it continues to inspire our efforts to reshape society, our fascination with "the future."

The recent assertion of Eastern ideas along the cutting edge of Western thought represents a reaction to extremes, but not a long-term retreat from action in this world. At best, the modern adventure is dialectical. It involves playing the edge between doing and not-doing, between effort and surrender, between selfish action and selfish quiescence.

Every great adventure needs a myth, a "story." Until fairly recently, the frontier has provided one great American myth, the rags-to-riches saga another. But we have become disillusioned with Paul Bunyan and the cowboy and with Horatio Alger as well, and it has sometimes seemed that we are a storyless people. Gradually, however, a new story is emerging, not just for Americans but perhaps for all the peoples of the planet. Charles Darwin wrote the first clear outline of this great story, though bits and pieces of it had emerged even earlier, in the works of Fichte, Schelling, Hegel, and Diderot.

The name of the story is evolution, the creation of ever-increasing complexity, order, and beauty against the field of entropy; evolution not only of biological forms but also of culture, society, and of consciousness. Nor is this evolutionary tendency limited to life on this planet; it seems to be the basic intention of the universe at large. Harvard astonomer David Layzer points out that "the universe is unfolding in time but not unraveling; on the contrary, it is becoming constantly more complex and richer in information."

This concept, which jibes well with experience and intuition, is relatively new for science. Influenced by the tidy, clockwork formulations of Sir Isaac Newton, many scientists of the past three centuries tended toward a kind of cosmic determinism that left little room for novelty and adventure. If only one could know all the facts at any given moment, such scientists as Pierre Simon de Laplace argued, the entire future would be clear. Not so, says Layzer. Since new information is constantly entering the picture, *not even a computer as complex as the universe itself could ever contain enough information to completely specify its own future states.*

"The present moment always contains an element of genuine novelty," Layzer writes, "and the future is never wholly predictable. Because biological processes also generate information and because consciousness enables us to experience those processes directly, the intuitive perception of the world as unfolding in time captures one of the most deep-seated properties of the universe."*

To visualize the difference between a static and an expanding, evolving universe, imagine balls in motion on a billiard table. If you video-taped a short segment of this motion, then entered it into a properly programmed computer along with other relevant information about weight, friction, and resilience, you could read out the position and momentum of the balls at any given moment in the future. But now imagine a situation in which new balls are being dropped on the table, new forces are being applied to the old balls, and new complexity being added to the table's shape and topography. In this case, the computer projection would be next to useless. In fact, that's the one prediction you could safely throw out.

This helps explain why the most distinguished economists are at sixes and sevens about future economic trends, why changes in population growth rate catch population experts off-balance, why new social trends are often obvious to thirteen-year-olds before they attract the attention of the best-informed social analysts, and in general why the most knowledgeably programmed computer projections into the future are sometimes accurate for two or three years, then go up in flames. In *Profiles of the Future,* Arthur C. Clarke compiles an impressive list of failed predictions in sci-

* Quotations from David Layzer, "The Arrow of Time," *Scientific American* (December 1975), pp. 56–69.

ence and technology: that heavier-than-air flight is scientifically impossible, that space travel is "utter bilge," that the energy in the atom can never be harnessed, and so on. Such predictions, generally made public only a few years before they are proved wrong, come not from the uninformed and the irresponsible but from the most highly respected and well-established people in their fields. In every case, novel and unexpected developments enter the picture, transforming the impossible into the obvious.

Perhaps the safest prediction we can make about the future is that it will surprise us. As Layzer points out, the present moment always contains an element of genuine novelty. The universe is continually at its work of restructuring itself at a higher, more complex, more elegant level. The novelty, the new, more complex order, doesn't emerge from the present in a steady stream, nor at all places at the same rate. It comes, as all things do, in rhythmic waves; there will always be times and places of scarcity and stagnation and retrogression. Still, the long-term direction is clear. The intention of the universe is evolution.

This being the case, adventure ultimately is not just possible but inevitable. Aware of it or not, each of us is involved in the grand enterprise of evolution. The new information being generated in each of our lives contributes inevitably to the ever-increasing complexity and richness of the universe. Our key choice is whether to become aware of and take responsibility for the power of our intentionality.

I've pointed out earlier that a truly centered person, experiencing that he or she is at the center of an expanding universe, will feel no compulsion to push and shove others out of the way. Similarly, a truly centered person, aware that he or she *intends* and *acts* in a universe of ever-increasing possibilities, will feel no need

157

to encroach upon or exploit others for selfish purposes, there being a practically unlimited supply of nonexploitative options from which to choose.

To put it another way, we are living essentially in a situation of plentitude, not scarcity—and our awareness of this fact can radically alter all of our experience and action. While there may be scarcity in the physical realm, even this is only temporary or apparent. The whole universe, including that which we call physical, can be treated in terms of information, and it is clear that information is constantly increasing. The supply is essentially unlimited, for we can always generate new information in the universe of the self—make new discoveries, new choices, put old things together in new ways—and this is bound to influence the future. To influence the future consciously, then, taking responsibility for the outcome of this influence, is to participate fully in the ultimate adventure.

Human intentionality expressed through science and technology has brought unprecedented material prosperity to the people of the advanced nations. Now many of these people are seeing limitations and dangers in further development strictly along the same lines. Science and technology will continue to play a great part in the evolutionary adventure, but perhaps in a different, less domineering way than before. More and more people are becoming interested not so much in robots or bionic organisms as in the remarkable evolutionary possibilities in their own bodies— which helps account for the explosive growth of the current fitness movement and the lively interest in holistic health. The body is indeed a universe in itself, and the effects of intentionality on this universe are swift and dramatic. Realizing this, people in increasing numbers are recapturing responsibility for their own bodies from the professionals. This means sixty-year-olds running sprints, hurdling, and pole-vaulting at performance levels that confound the norms of the medical profession. It means ordinary citizens using the powers of the placebo effect—without placebos or doctors' prescriptions—to achieve feats of self-healing and positive bodily transformation that bring the yogic legends to mind.

Personal concern can lead directly to social concern. Indeed, once responsibility and intentionality have been applied successfully to the body, they are likely to be applied to the body politic as well—and in a way that, again, will surprise the experts.

We can see one such application on a large scale in the Hunger Project, initiated by *est* founder Werner Erhard in February 1977. Like similar enterprises, this project encourages its participants, now more than 150,000 strong, to fast, contribute time and money, influence public policy, support antistarvation organizations, and work directly on the problem in the field. But the main idea behind the project—and this gives it genuine novelty—lies in creating a new context for the problem that will permit human intentionality to go to work on its solution. Just as Arnold Schwarzenegger could "see" himself as Mr. Universe in a way that was vivid and absolutely real years before he won the title, so the Hunger Project enrollees are encouraged to "create" or "visualize" a planet totally without hunger and starvation by 1997, to make this vision vivid and absolutely real at the deepest level of their being, and then to act in terms of the context that is expressed in that vision. Each enrollee, as strange as it might at first seem, is asked to take personal responsibility for the end of hunger.

At first glance, this might seem to smack of voodoo or simply to constitute a kind of consciousness-raising. Actually, it doesn't necessarily entail anything paranormal, and "consciousness-raising" is only part of the story. When the end of hunger and starvation by 1997 is *made real,* then everything that happens around the problem is seen in a new perspective. Carelessly held assumptions—of scarcity, of inevitability, of lack of viable solutions—fall away. Unexpected and novel measures for action suggest themselves. Nor is there any longer the need to waste time and energy, as is now so often the case, fighting for any one position on the problem as opposed to other positions. Every position, even the position that the problem has no solution, can be seen as having a certain role in the new context.

Though it involves thousands of people, the Hunger Project is by no means a mass movement, but rather an alignment of individuals. Each individual is seen as a whole, a context which includes "the end of hunger and starvation by 1997" as an integral part. The project is personal, but not in the sense of being private or quiescent. In a few short months, it has involved large-scale meetings in eleven U.S. cities as well as many smaller meetings, conferences with other hunger organizations, with scientists, with legislative groups, with high officials at the White House, with Prime Minister Desai of India and with leaders of other underdeveloped nations. According to Roy Prosterman, a widely respected expert on world hunger, "I've seen more interest and action on hunger in Washington in the five months since the Hunger Project began than during the previous ten years."

It will be instructive to follow the project during the next several years. Intentionality is involved in all human enterprises large and small, but rarely is it expressed so explicitly or applied so directly and unequivocally to a specific problem as it is in this case. Since intentionality works on the level of identity

THE INTENTION OF THE UNIVERSE

rather than ego, and identity does not involve itself in
credit or blame, a main premise of the project is that
it must not, cannot, take credit for the end of hunger
or for any of the positive steps along the way. Still, we
have seen in earlier examples how a strong intentionality seems to exert a mutual influence on its surroundings, and there have been several "coinci-
dences" of this type since the project began.

During recent decades, for example, expert opinion
on world hunger has tended to take a characteristically
gloomy tone; solving the problem has generally been
deemed "impossible." Five months after the Hunger
Project was initiated, however, a panel of experts convened by the Research Council of the National Academy of Sciences released a 192-page report on the
subject that was notable for its optimism. "If there is
the political will in this country and abroad," the report stated, "it should be possible to overcome the
worst aspects of widespread hunger and malnutrition
within one generation. . . . We find these prospects
exciting and worthy of strong national efforts, and we
believe that a latent political will now exists in numerous countries which could be mobilized in a mutually
supporting fashion to commence and support such
efforts."*

There have been other interesting developments.
On February 3, 1978, President Carter announced
that he was setting up a Commission on World Hun-
ger, the first such commission in American history.
This announcement followed intensive lobbying by
Hunger Project people at the White House. And on
February 14, in a dramatic piece of synchronicity,

* *World Food and Nutrition Study,* 1977, available from National Academy
of Sciences, Printing and Publishing Office, 2101 Constitution Ave , N.W.,
Washington, D.C. 20418 ($6.75).

United Nations officials announced that, for the first time in 3 million years of human existence, world population growth has started leveling off.

Another coincidental development, made public several months after the project began, borders on the miraculous. Searching for an alternative to chemical fertilizers (which require oil or natural gas), a team of Michigan State University researchers led by Stanley K. Ries extracted a simple alcohol compound from alfalfa. The compound, called triacontanol, contains the same elements, only more of them, as the ordinary alcohol that people drink in liquor. The researchers then began varying the amount of triacontanol in water to learn just what concentration most effectively stimulated plant growth. To their amazement, they found that *one* part of triacontanol to *a hundred million* parts of water was by far the best.

Just what was going on here? Had the research group discovered a plant placebo, or was it a case of vegetable homeopathy? Whatever the explanation, the drastically diluted substance, sprayed on ten different vegetable crops, first in rigidly controlled greenhouse experiments and then in field tests, boosted food output in an extraordinary manner. Carrots treated with the new substance grew bigger and fatter and produced up to 21 percent more foodstuff than did untreated carrots in the same soil. Asparagus plants produced 35 to 60 percent more. Sweet corn yield increased by nearly 25 percent, while tomato yields have been as much as 30 percent higher than those of plants fertilized normally.*

As for adverse side effects, Ries points out that tria-

---

* Stanley K. Ries, Terry L. Rickman, and Violet F. Wert, "Growth and Yield of Crops Treated with Triacontanol," *Michigan Agricultural Experiment Station Journal,* (Jan. 3, 1978).

contanol is a natural constituent of many leafy vegetables and of potatoes as well. "Anyone eating a normal meal," he said, "is eating enough triacontanol to spray an acre of vegetables."

It sounds almost too good to be true, and perhaps there will turn out to be a catch somewhere along the line. But American Cyanamid Company has already taken an option for an exclusive license from Michigan State to make and market the magical substance. If this ephemeral alcohol spray does prove out in its final large-scale tests on many continents, it could produce a revolution in food production and go a long way in the fight against world hunger. Its discovery also illustrates how the truly novel and unexpected can burst into existence in this universe of ever-increasing complexity and richness of information, and how such new developments must inevitably confound every expert extrapolation into the future.

Other surprising events have occurred since the birth of the Hunger Project. Does this mean that the project *caused* the miracles? Not at all. Or maybe it was the National Academy of Sciences panel that caused Hunger Project? Again, not the point. As we've seen earlier, the whole concept of causation is less than useful in these cases. What can be said is that intentionality, like some wonderful kind of psychic gravity, exerts a mutual influence, gathering in, weaving together the stuff of existence in new form and sequence to further the evolutionary adventure.

Even after we realize that we live in a universe of ever-expanding possibilities, it's sometimes hard to see how "mere" intentionality can affect the ponderous realm of matter and break the chains of history. How much easier it is to take refuge in our essential helplessness than to take responsibility for our godlike powers! Here, too, our picture of reality shapes our attitude and actions. When we see the body as a structure of heavy, intractable meat and bones, all we can do is cut it or drug it or otherwise manipulate it from the outside. But let us create a different picture and see the body as rhythm and emptiness, as elegant, ethereal fields of waves joined in innumerable feedback loops to maintain a relatively stable structure. Thoughts and feelings can set off sympathetic vibrations in this structure; the deeper vibration that I have called intentionality can produce transformations in days, hours, sometimes minutes. No longer are we concerned with questions of "inside" or "outside," for these limiting categories do not exist in the realm of pulsing waves. Each body completes the universe and is involved in the universal evolution.

In the same way, we can picture history, not in terms of chains, but rather in terms of cascading waves unfolding in time, producing new information, new options—"miracles," if you wish. This new information, whether it be a novel arrangement in the DNA molecule to produce a new species or a novel arrangement of language to produce a new idea, is what ultimately rules. Meat and bones, cathedrals, species, the tread of mighty armies, all are subsumed in rhythm, and a rearrangement of vocal sounds or pen scratches on paper can bring an empire down.

In such a universe, how can we think of ourselves as helpless?

ELEVEN
"Life Cannot Be Fooled"

There are times when the ideas that have emerged in this book seem strange to me, paradoxical, an affront to the common sense passed down from generation to generation in this society. Let's list them now in some kind of order, nine theses:

1. That we are composed of waves no less than of the stuff we call "solid."

2. That what we call objects and events are primarily the precipitates from the relationship of these waves.

3. That each of us has an *identity* that is unique in all the universe, and that this identity is expressed as a distinctive wave function.

4. That each of us is also, paradoxically, a *holoid* of the universe, containing universal information—past, present, and some of the future.

5. That knowledge of the future fades away from us simply because the universe, by its very nature, is con-

stantly creating unforeseeable new information, genuine novelty, and that the destiny of the new information is the evolution of higher forms.

6. That each of us is, in essence, a *context,* a weaving together of universal information from a particular point of view.

7. That what can be called "perfect rhythm" exists at all times in the paradoxical interplay, the silent pulse, between identity and holonomy within the context of each of us, and that, beyond custom, language, and ego, we can directly experience this perfect rhythm.

8. That *intentionality,* the vector of identity, is an essential element in the universe that is each of us; and that it is possible, through intentionality, to influence this universe in extraordinary ways.

9. That, in potentia, we know everything.

That we can know everything seems perhaps most difficult to swallow. How can this be, when we are so enmeshed in ineptitude, so smothered in ignorance? But here it's good to emphasize that what we are *conscious* of is as small as a pea in the ocean of the self. What we can put into words, shape into images and concepts, is indeed minuscule. As Michael Polanyi has pointed out, we know more than we can say, and we can see this quite clearly in our human relationships, where there are no secrets, only the conspiracies of willful ignorance. No husband has ever really deceived his wife. That which is most deeply buried reverberates most painfully in her every cell and in his, compromising every word and look and caress between the two of them. In spite of the cleverest deception, she knows, and he knows she knows. The dying patient is not fooled by the reassurances of physician and family. Unacknowledged, death dominates all that passes for conversation.

With what devastating force every concealment makes itself felt! In *The Devil's Share,* Denis de Rougemont relates the brief history of a correct couple, a Mr. and Mrs. who are scrupulous in maintaining an air of peace in their home, never letting their maid or children witness their fights. When their daughter, Mary, develops nervous problems, the wife tells a psychiatrist that her husband's genes are probably to blame.

"Between ourselves, one of his aunts died in an asylum. Sometimes it shows in him. Just yesterday, to give an example, we had barely come up to our room when he got into a fury because I asked him to put out a light that was hurting my eyes. He threw it on the floor and made one of those scenes! I wept all night on a couch in the living room . . ."

"Madame," said the doctor, "your children know all that."

"Impossible, Doctor. Knowing my husband, I have them sleep at the other end of the apartment."

"I tell you they know everything without hearing anything. What they hear best is everything you don't say, when you are before them at the table, so polite. Little Mary is not mad, but how could the nerves of a child stand the sound and fury which are unleashed in your conjugal silences?"

A man can know the world, the Chinese sage, Lao Tzu, tells us, without ever having left his room. Kant traveled no more than fifty miles from his birthplace and created a vast landscape of thought. Bach's life was restricted not only geographically but culturally; his ignorance was reflected in his ugly scrawl, his ungrammatical German, his confused diction. Yet he took the measure of the heavens and reproduced in sound the grandest cycles of the cosmos. Emily Dickinson rarely left her house. "I never saw a moor," she wrote, "I never saw the sea;/Yet know I how the heather looks,/And what a wave must be." The brilliant and prolific contemporary fictionist, Joyce Carol Oates, writes compellingly of gluttons and torturers and rapists. This slim, shy, retiring woman told me she simply serves as a medium who, like all of us, she feels, has access to inexhaustible sources of knowledge.

We might be tempted to explain all this in merely artistic and intellectual terms. But there's more to it than that. For example, during our lives, most of us have heard of some method of torture so ingenious, so improbable, so hideous that it would seem to be totally outside our own experience and quite beyond the powers of our imagination. Yet the shock we feel is one of recognition. A man's belly is cut open, his intestines uncoiled and burned, he is forced to watch the spectacle. Yes, it is horrible beyond all belief—and somehow familiar. At the heart of it, it is we who have devised that torture.

Let's be straight about it. Life is not "nice" or "correct." If we know everything, we know horrors beyond the power of words to express. If we "are" the universe, we can't plead our innocence to the most heinous of crimes. We have ravaged and pillaged with Attila the Hun. Clothed in flayed and rotting human skin, we have sacrificed young children, vying with the other priests of Tenochtitlán to see who could make them cry the most pitiably. Adolf Hitler is no stranger at our pleasant suburban home, but rather a familiar guest. We know his predilections well, his insane rages and bloody dreams, for they are ours, too. He will hold us forever in fascination, in fact, until we own all of this, until we admit that the Hitler in us is also a charming conversationalist, a considerate lover, a delightful playmate with visiting children. We will lay his persistent corpse to rest in the history books only when we acknowledge our desire for just the kind of Dionysian ecstasy he promulgated and experienced: the cry of *"Seig Heil!"* from a hundred thousand throats.

"I am a man: nothing that affects mankind do I think a matter of unconcern to me," the playwright Terence wrote in the second century B.C.; the idea, you see, is not a new one. But what I am proposing supersedes the specifically human. My thesis requires that we have implicit knowledge of life in distant galaxies, and suggests that the fantasies of science fiction may turn out as reasonably accurate maps of the future. I'm suggesting, as well, that all the cosmic patterns, the "secrets of the universe," are embedded in our muscles and bones and cells, in whatever makes up the context of the self, and that these patterns must in some way influence the way we think and feel and act.

For example, the notion of the hologram, the very idea of a fragment of a picture that contains all of the picture, offends our common sense. At the same time, it strikes us with a sudden, unexpected thrill, that sense of recognition again. There have been other extraordinary discoveries in recent years, none more outlandish than that of the black hole. This bizarre entity is a product of stellar evolution; when a star of a certain mass reaches a certain density, its gravity crushes its own matter out of existence, leaving only an infinitesimal "singularity," like the smile of the Cheshire cat, marking the spot in space at which the star disappeared. Though the matter is gone, the gravity and momentum linger on, as the strange nonobject continues its cosmic journey, influencing other objects as it passes. The singularity is surrounded by a sphere of some ten to twenty miles in diameter—the black hole itself. So powerful is the gravity in the hole that whatever enters it can never get out; even light will be sucked down into the singularity to disappear or, in the minds of some scientists, to reemerge in another universe. The whole notion is preposterous, yet it seems to resonate with certain human experiences. Surely there are black holes in our personal context; on some days it seems that way in any case. So while astronomers continue their search for black holes with radio telescopes and cosmic-ray gauges and gravity-wave detectors, we can look inside for a different kind of confirmation.

Yes, at some level we do "know everything." We can *feel* the catastrophic gravitational collapse that creates a black hole in space. We can *feel* the symmetry of ascending frequencies that creates the electromagnetic spectrum and the musical scales. We can *feel* a seedling's urge to grow, the suffering of a starving child in India, the sound of music on a distant planet. Understood this way, the conversational entrainment described in Chapter 2 is no longer a mystery. At the depths of our being we know, even a baby knows, the rhythm of the speaker's voice; this secret is revealed in the subtle dance of the body.

It is all available in the vast, mostly unilluminated context that we have called the personal body, mind, and spirit. The act of illumination involves not going out and finding what isn't there, but simply shining the light of consciousness on what *is*. Expanding the area of illumination is an exhilarating and often painful process, for the limited circuitry of our conscious awareness can easily be overloaded. Every new vista shocks us with its terrors along with beauty newly realized. And it is often difficult for our primitive intellects to reconcile the seeming opposites out of which is woven the ultimate harmony of the universe.

In spite of our qualms, however, the illuminated area widens, the beams of human thought become more powerful. The equations of modern physical theory cast new light on the ancient intuition that all the parts of the universe contain the whole. The exotic new logic of quantum physics suggests that we exist as a locus of waves that spreads its influence to the ends of space and time. New discoveries and new formulations surprise us—and the surprise has a ring of familiarity about it.

And then there is genuine novelty, not the discovery of something that already exists, but the emergence of the truly unexpected, a gift from the universe of some new information, some new configuration. These gifts sometimes manifest themselves in the shape of a Christ, a Buddha, or a St. Francis, a brief human life that can change the course of history. Perhaps we have had such gifts in our own time.

The preacher kept coming back to the same phrase, not so much spoken as intoned. Again and again he would lead his congregation to the depths of despair with chronicles of tyranny and injustice throughout human history. Then once more he would sing it out: *"But life cannot be fooled."* And he would tell us of the tyrant overthrown, injustice overwhelmed by the intrinsically redeeming forces of existence. His voice was like music, the oft-repeated phrase a triumphant refrain *("Life cannot be fooled")*, bringing forth soulful cries from the congregation: *"Yes Lord!"* *"Praise Jesus!"* *"Say it clear!"*

It was the first Sunday in February, 1961, in Atlanta, Georgia, and the assistant pastor of the Ebenezer Baptist Church, a young man named Martin Luther King, Jr., was delivering the sermon. I looked around at a sea of black and brown faces; Cal Bernstein and I seemed to be the only whites there. We had come south on assignment for *Look* magazine to do an article on the Atlanta sit-ins and boycott, which happened to be the focal point of Civil Rights activity that winter. Atlanta schools were to be desegregated the following fall; now there were lunch counters to integrate, demonstrations to be mounted, and a boycott to be enforced against Rich's, the largest department store in the South. The Civil Rights movement was well underway by then. Atlanta, the pivotal city of the Deep South, was to be an important test of King's nonviolent tactics. It was also to be an occasion of consolidation and celebration, for what had seemed fantastical and beyond all realistic hope was indeed turning into reality.

Every human being is blessed and cursed with an automatic mechanism that makes the improbable, even the miraculous, once accomplished, seem inevitable and commonplace. Now that every public facility in the South is integrated, now that black and white laborers sit together at lunch counters on the lonely back roads of Mississippi, now that black and white children share the same classrooms, water fountains, rest rooms, and swimming pools, now that blacks flock to the polling places and major Southern cities have black mayors and there are Southern blacks in the U.S. Congress and one of Martin Luther King's top lieutenants is the American ambassador to the United Nations—now that all this has happened it's easy to shrug and say, "Yes, but look at the job situation," or "Yes, but have hearts really changed?" or any number of other "yes buts."

Fully to realize the magnitude of the miracle, you would have to go back to the 1930s when I was growing up in the Deep South, or even to the time of the Supreme Court desegregation decision in 1954, when the vast majority of Southern white men stated loudly and repeatedly—and believed—that they would die before allowing a single black child to go to school with one of their daughters. The most respected social observers of that time, Northern as well as Southern, foresaw a lengthy struggle, many decades long, involving endless legal maneuvering, open defiance, and, in case the issue were joined, massive bloodshed. No change is more painful than one that involves facing up to and relinquishing a deeply rationalized immorality. Clearly, the South said, "Never!"—as the news media of the times headlined. Anyone who stood up in 1954 and predicted that things would happen just exactly as they have happened would have been called a fool.

As we've seen, however, the universe is in the business of delivering up the unpredictable. No one in 1954 could have predicted the real meaning of that moment to come on December 1, 1955, when a soft-spoken black woman named Rosa Parks would refuse to move to the back of a Montgomery bus. No one could have predicted that the young Baptist preacher who took up her cause there on the plains of southern Alabama would have been schooled in the philosophy and tactics of *satyagraha,* the nonviolent "truth force" of Mohandas Gandhi. No one could have predicted that the new communication medium of national television would present the battle of *satyagraha* and soul versus racist hate with such electrifying effects.

In Atlanta, it all came together. I had covered and would cover Civil Rights battles from Little Rock to Ole Miss and Selma, and would remember Atlanta not so much for its tensions as for unexpected moments of laughter and the richness of its music. Hearing *We*

Shall Overcome for the first time during the Atlanta
campaign, I hurriedly wrote out the musical notations,
not realizing that that simple melody would soon resonate all around the world, to become the anthem of
oppressed people everywhere. But those weeks remain vivid in my memory primarily for the presence of
Martin Luther King, Jr., and, most of all, for his sermon at Ebenezer Baptist Church.

Two years were to pass before King was to be
awarded the Nobel Peace Prize, and on this Sunday
morning in 1961 he was speaking only for his own
congregation of mostly middle-class blacks in a Southern city in the United States. Yet it seemed to me he
was already speaking for all of humankind. His message was simple. Creation, justice, harmony—these
are the ultimate laws of the universe. Sometimes injustice arises to challenge heaven, grows strong and
seemingly irresistible. But every Napoleon has his
Waterloo, every Hitler his fiery bunker in Berlin. "Life
cannot be fooled."

King approached his subject through four separate
modes of discourse. He spoke philosophically, histori-
cally, religiously, and emotionally, switching from one
mode to another with an inexorable rhythm. He
quoted Nietzsche, Kierkegaard, Schopenhauer, and
Gandhi. He sketched the story of Israel's flight from
Egypt in a dazzling two-minute vignette. He drew a
picture of the African slave ships so compelling that,
had he continued another minute, he would have had
the entire congregation crying aloud. Then, unwilling
to let us lose ourselves in emotion, he brought us back
with a telling point from the *Nichomachean Ethics* and
reminded us of our duty to live according to universal
law even if it meant suffering and death.

During this incredible performance, Martin Luther
King, Sr., a powerful presence in a great oak chair

above the altar, would sometimes hit the arm of the
chair and laugh aloud with wonder. At one point, a
voice from the back of the church rang out, "Preach!
Preach! That man can preach!" Cal Bernstein and I,
like everyone else, were swaying in rhythm with King's
words, and, though the church was cool, our faces
were wet with perspiration and tears. Now and then I
could hear Cal saying, "Yes, Lord!" along with the
rest of the congregation, something totally unlike him
yet totally right; I heard myself doing the same. Once,
Cal leaned over to me and whispered, "Too bad this
isn't being recorded. Everyone in the country should
hear it." "Maybe we're not ready for it," I whispered
back. "It's too beautiful."

That's how it was for me: nearly too beautiful to
bear. The church became a mythic place (Mount Ara-
rat on Auburn Avenue). The passing moment seemed
balanced on the knife-edge of history and time. Colors
were more vivid than usual. Every sound was crisp and
eternal. In his dark robe, King stood before us like a
prophet. His faintly oriental eyes glistened with ancient knowledge and he spoke of human destiny with
the vehemence and majesty of music.

King led us toward a consummation by outlining
the saga of slavery and liberation, suffering and tri-
umph in America. Somehow, "the wonderful Negro
people" had prevailed over injustice, transmuting bit-
terness and hate to compassion and love in a crucible
of deep humanity that might now constitute a great
redeeming force in the world. The current Civil Rights
movement, he said, was offering us a chance to join
our lives with the flow of the universe, at the heart of
which could be found forever the creative power of
love. The mighty would fall. The songs of slaves and
sharecroppers would rise to the heights of earth and
heaven. Even the most glittering tyrant eventually

175

would learn how futile it is to stand on the beaches of history and try to hold back the tide.

The sermon ended, the prayer, the benediction. Unrecorded, the precise words slipped away, but the experience is still with me. King's voice had touched my every muscle, every cell, and I was never again to be quite the same. For what had entered me was not just words, but a majestic rhythm; the indomitable, ever-surprising music of the universe itself. *"Life cannot be fooled."*

Let us now, if only for a few moments, accept the incredible: that beneath the world we take as real there lies a deeper reality, another world in which our common notions of time and space and momentum are overthrown, in which our careless faith in the impossible begins to lose its age-old grip. A strange place is this world of the new physicists, a world of ultimate connectedness, where consciousness—or observership, as John Wheeler calls it—coexisted with the creation, and where it might be said that the vastness of space, the nuclear conflagration of stars, the explosions of galaxies are simply mechanisms for producing that first glimmer of awareness in your baby's eyes.*

How are we to live in such a world? A snarling cab driver cuts you off. Your temper rises. But wait. . . . The cab driver is a weaving together of the whole universe from a particular and unique point of view. Beneath the twisted face and clenched fist, there beats the pulse of perfect rhythm, silent and unacknowledged, but still present. Let us say that the snarl springs from unfulfilled potential; perhaps it is a cry

* John Archibald Wheeler, "Genesis and Observership," *Foundational Problems in the Special Sciences* Butts and Hintikka, eds., (Dordrecht, Holland: 1977), pp. 3–33.

for help. No need, in any case, to add to the static and confusion by snarling back.

At the same time, there is no need to be passive, a victim of circumstance. The view of reality sketched out in this book suggests that each of us possesses practically infinite amounts of unused power and knowledge; that, no matter how boxed in we might imagine ourselves to be, alternatives are possible; and that we are not and can never be mere observers, but must always play the role of active participants at the feast of life.

It is fascinating to me that the physical scientists give us these insights just at the moment when the world is ripe for transformation. Obviously, the time is past for us to consider the self as doggedly separate from and opposed to the rest of existence. This tragic dichotomy between self and other, along with glorification of competition and "winning" and general selfishness, might have worked in the days of the frontier, but in recent years it has led unerringly to Vietnam and Watergate and the energy, ecology, and economic crises. Simple historical logic—no less than the theses offered here—urges that we transform our values and behavior.

A world of connectedness, potential, and evolution turns us toward a vivid sense of community along with the acceptance of personal responsibility; toward a de-emphasis on competing and winning along with a re-emphasis on participating and experiencing; from aggression toward gentleness and enjoyment; from dominance of nature to blending with nature; from exponential growth in production and consumption to a more moderate, more ecological standard of living along with a powerful intentionality; toward social justice throughout the world.

Perhaps more than anything else we can say in

words, the new world offers us—whenever we are ready to take it—the thrill of novelty and surprise. Can we call it Transformation?

Ceaselessly pulsing, the universe expands. Stars burn out; galaxies spin more slowly. In terms of physical energy in any given area, the laws of entropy, of increasing decay and disorder, reign. Yet, paradoxically, the universe as a whole is constantly generating new information, new order, new complexity. This evolution of higher form does not proceed at an unvarying rate, but in pulsing waves. A million fruit flies die to bring forth a single viable new strain. A new culture or a new musical form comes to life out of the long downward spiral of the old. Catastrophe seems to loom ahead of us, and the voices of cynicism and despair grow loud in the silence of vision. But the cynics see only one side of the wave. On the other side is some unexpected configuration of forces, some new Martin Luther King, Jr., who can transform the social order of centuries in a few short years; or some new invention like the transistor, which is even now creating a worldwide computerized communications network and transforming the social order of millennia.

In terms of game theory, we might say the universe is so constituted as to maximize the play. The best games are not those in which all goes smoothly and steadily toward a certain conclusion, but those in which the outcome is always in doubt. In baseball, for example, we could move second base six feet farther from first base and make base stealing practically impossible. Or we could move the bases six feet closer and make nearly every attempted steal successful. Either of these changes would make things smoother and more predictable—and would spoil the game. As it is, tension is high when a player is on first base. The game is at its best. Similarly, the geometry of life is designed to keep us at the point of maximum tension between certainty and uncertainty, order and chaos. Every important call is a close one. We survive and evolve by the skin of our teeth. We really wouldn't want it any other way

The game is evolution and each of us is fully involved. Through our intentionality, we can change our bodies and the body politic in surprisingly effective and dramatic ways. We are doing so all the time, whether we acknowledge it or not. At the depths of our being, we do know everything. No matter how limited our situation seems, even in a prison cell or on our final plunge to death, we are rich in options as to how we experience the moment. Just by transforming the moment, to some extent we transform all of existence.

And this, perhaps, is most difficult for us to accept: not that we share the predilections of a Hitler, but that we also are like gods. I'll not press this idea upon you. Knowing my own limitations, seeing the sadness and waste of potential all around me, I know very well how frightening and painful it is to realize the transformative powers we finally must know we have. The cynic's role has a long and respectable history. I'll not try to deny you its comfort and safety.

But there are shimmering moments on the mountaintop when the chaos dissolves in the sweetness of reflected sunlight, moments of perfect rhythm. There are interludes in the evening by the fire with music filling the room, music that joins the life of the body with the rhythms of the universe. Seeming contradictions are reconciled in that vibrancy, not through the denial of death, violence, suffering, and injustice, but through the vital, pulsing force that makes harmony of opposites. During those moments, all that *is* is music,

music heard and unheard, in which negation and self-deception are impossible. And then all the connections are clear and I know that Martin Luther King was right.

Life cannot be fooled.

APPENDIX

Toward the Experience
of Perfect Rhythm

You should be balanced and centered before beginning any of the exercises in this appendix. The best way to start out is to have one person read the following instructions aloud to you (or to a group of people). This will take ten to fifteen minutes. Later, after practice, you will probably be able to balance and center yourself in a matter of seconds, with no outside help.

Note that three dots indicate a definite pause. Before proceeding, the person reading the instructions should check on his or her own state and also sense whether those following the instructions are ready to go on to the next step. Instructions should be read in a clear, unhurried voice:

"Please stand with your feet slightly farther apart than your shoulders, eyes open, knees not locked and not bent, trunk upright, arms hanging relaxed by your side. . . . Now take the index finger of your right hand and touch it to a spot an inch or two below your navel. Press in firmly, toward the center of your abdomen. This is your *center,* what the Japanese call the *hara.*

This is the physical center of gravity of your body. Now drop your hand to your side. Let air enter you through your nostrils, and let it move downward through your body as if it were going directly to your center. Let your abdomen expand from the center outward to the front, to the rear, to the sides of the pelvis, to the floor of the pelvis. . . .

"As your breathing continues in a relaxed manner, lift your arms in front of you, with the wrists entirely limp. Shake them so hard that your whole body vibrates. . . . Thank you. Now lower your arms slowly to your sides and when your hands touch your legs or any part of your clothing, let that be an automatic levitation switch that causes your arms to float upward. Let your arms float upward directly in front of you, with the hands hanging limply. As the arms rise, lower your body by bending the knees slightly. Keep the trunk upright. The arms rise as if floating up in warm, salty water. When your arms reach the horizontal, thrust the palms forward into the position you would use if gently pushing a ball on the surface of the water. Shoulders relaxed. Now sweep the arms from side to side as if you can sense or 'see' through your open palms. How would it be if you could sense things through your palms? . . . Now shake out your hands and repeat the process. When your hands touch any part of your body or clothing, that's an automatic levitation switch. Let your arms levitate by floating up in front of you as if in warm, still water. As the arms rise, the body lowers. Knees bent, trunk upright. Now thrust your palms forward and sense the world through your palms, sweeping them from side to side. What would it be like if you could sense a person near you through your palms? Give it a try. How would it feel if you could? . . .

"All right. Drop your hands, and this time leave

them hanging by your sides, naturally, in a totally relaxed manner. Please close your eyes. Knees not locked and not bent. Now check and see if your weight is balanced evenly between your right and left foot. Shift slightly from side to side, fine tuning your body. . . . Now check and see if your weight is balanced evenly between the heels and balls of your feet . . . knees not locked and not bent. . . . Please leave your eyes closed and shift to a more comfortable position any time you wish. . . . Now move your head forward and back to find the point at which it balances on your spine with the least muscular effort. . . . Take a moment now to relax your jaw . . . your tongue . . . the muscles around your eyes, your forehead, temples. . . . Thank you. Please turn your attention for a moment to your shoulders. Let them melt downward, like soft, warm chocolate. With each outgoing breath let them melt a little farther. . . . And now the lower pelvic region. Let that relax also. Release all tension. With each outgoing breath, let go a bit more. . . .

"And now please consider the back half of your body. What if you could sense what is behind you? What if you had sensors, eyes in the small of your back . . . at the back of your knees . . . at the back of your heels? . . . What would the world look like from that angle? . . . Or from the back of the neck? . . .

"Now send a beam of awareness throughout your entire body, seeking out any area that might be tense or rigid or numb. Just illuminate that area, focus on it. Awareness alone often takes care of these problems. . . .

"Once more, concentrate on your breathing. . . . Be aware of the rhythm. Now, in rhythm with one of the incoming breaths, let your eyes open. . . . With eyes softly open, walk around slowly, maintaining the relaxed state you've achieved. . . . Let your physical center be your center of awareness. . . . Can you entertain the idea of this center as the center of the universe? . . . What would happen to your own self-aware consciousness if your center were also the center of the universe? . . .

"As you go through the rest of the day, you might recreate the centering and balancing process at various times. Please bear in mind that the body is a metaphor for everything else. Your relationships, your work, your life itself can be centered and balanced. And when you're knocked off center in one way or another, there is always the possibility—if you can stay fully aware—of returning to the balanced and centered condition at an even deeper level."

RETURNING TO CENTER
There are numerous ways of practicing this skill. Among them:

• Stand with eyes closed, balance and center yourself, then lean over from the waist. When you are accustomed to this position, straighten up rather suddenly and open your eyes. Experience your sense of disorientation fully; don't deny the reality. Then settle down into a centered state. Be aware of what happens during this process. Does the condition of being centered and balanced seem somehow more powerful or radiant after having been momentarily lost?

• Go through your balancing and centering procedure standing, with eyes open. Leaving the eyes open, spin several times to the left, then to the right—just enough to become slightly dizzy. Then stop spinning and return to center, with some increased awareness of the soles of the feet. Again, be aware of the process.

• With eyes open, balance and center yourself in a wide stance, arms held a few inches out from the sides. Have someone walk up silently behind you and grab one of your arms at the wrist, with enough impact to cause you to start. Be aware of how you have been knocked off-center. Describe it specifically to yourself. (For example, "I felt a rush of energy like an electric current shoot up my left arm," or "My heart seemed to jump up into my throat.") Then return to center, lowering your body by bending the knees slightly. *Use the energy of the sudden shock to aid in the centering process. Let your whole being expand.* Consider the possibility that a sudden shock experienced in this manner can be taken as a gift.

SOFT EYES

For most of us in this culture, normal vision entails focusing the eyes on specific formal entities, giving them shape, cultural meaning, and name. This kind of seeing, which I'm identifying with the term *hard eyes,* is basically analytical, having the effect of separating figures from the ground in which they may be said to exist—creating "objects" and drawing sharp edges between these objects. Seeing with hard eyes is a positive act; it requires reaching out into the world. With hard eyes we can read the fine print.

Hard eyes are appropriate for many, but not all, situations. The visual mode I'm calling *soft eyes* provides an alternative. This mode is receptive rather than positive, synthesizing rather than analytical. It involves letting the visual world come in rather than reaching out to bring it in. With soft eyes we tend to perceive a whole field of vision in terms of the energy and motion that make it up, rather than perceiving the collection of discrete objects that exist within it. There is less than the usual distinction between figure and ground. With soft eyes, peripheral vision is enhanced, the depth of field appears to be greater, and colors seem remarkably vivid.

Using soft eyes entails not just adopting an alternative visual mode, but also entering an altered state of being. Once you've mastered the art of soft eyes, this state can be achieved in a split second. The following detailed instructions are offered by way of introduction:

Begin by standing, with eyes closed, in the balanced and centered state previously described. Here, it's especially important that your shoulders and lower pelvic region be relaxed; soft eyes are practically impossible to achieve with tense shoulder muscles or a tightened pelvic region. Breathing, as always, should be spontaneous.

With the pads of the fingers of both hands, very gently massage both eyeballs through the closed eyelids until the eyeballs seem to soften, if only slightly. Then let your hands drop to your sides, and take a moment to check that the arms are totally relaxed. Become aware of your breathing. On the second or third incoming breath let the eyes open and let the world come in. Do not *stare* or try to keep from blinking. Do not *reach out* with your eyes to focus on any object or any point in the visual field. This is not a matter of throwing the eyes out of focus. You might say the eyes are focused on infinity, but even that wouldn't be entirely accurate. It would come closer to suggest that the eyes are focused on nothing in particular. They are simply open.

Now let yourself become aware of the entire visual field, giving no part of it any more importance than any other part. Let any movement or shape or color be an integral part of the whole field, related to everything else in it.

Leaving the feet planted firmly on the floor, rotate the body to the left and right from the hips, letting the arms swing freely. In this movement, soft eyes can sweep the entire sphere of vision and effortlessly compute the relationship of everything in view. You might note any increase in the depth of field and the intensity of colors.

After this, you might try walking around, practicing the use of soft eyes with a moving field. Note that you can shift from hard eyes to soft eyes instantaneously. The ability to make this shift is a key element in most of the exercises in this appendix.

There are also many practical applications of the use of soft eyes, especially in sports. A football quarterback, for example, must be able to perceive the entire flow of a pass pattern rather than fixing his eyes on one receiver. Many excellent passers fail to make the grade in professional football because when their primary receiver is covered they have difficulty picking out secondary receivers. The use of soft eyes can remedy this problem. Players other than the quarterback, especially running backs, defensive backs, and members of the special teams, also need to use soft eyes, as do participants in basketball, soccer, hockey, and all sports that involve the free-flowing movement of a number of players. The next time you watch a basketball game, note the expressions on the players' faces: that relaxed, seemingly vacant look in the midst of hectic action. Nobody has taught such players the art of soft eyes. It's simply that those who have learned it intuitively are the ones who, if otherwise properly endowed and motivated, have become the top performers.

My own practice of soft eyes is associated with the martial art of aikido, which emphasizes multiple attack. When three or more agile aikidoists are rushing in at full tilt, soft eyes are a necessity; there's no time to read the fine print. I've learned as well that soft eyes are appropriate for almost all aikido situations, even for slow-motion partner practice. With this mode of seeing and being, coordination is easier, body movements are smoother. Soft eyes are also helpful in many of the situations of daily life—driving your car through a tight spot, for example, or even walking along a crowded sidewalk. In nature, the movements of birds and other wild creatures can be picked up more readily with soft, rather than hard, eyes. All intuitive powers seem to be enhanced.

Shifting from hard eyes to soft eyes apparently involves a change in brain wave pattern. I was the subject of neurological experiments carried out at the laboratory of Dr. Tod Mikuriya, a Berkeley, California, specialist in psychosomatic medicine who is currently conducting research in biofeedback. With hard eyes, both the right and left hemispheres of the visual cortex of my brain pulsed at sixteen cycles a second. This is a typical beta-wave state, denoting a normal condition of attentiveness. When I reported going into the soft-eyed mode, the right hemisphere continued pulsing at sixteen cycles a second. The left hemisphere, however, dropped to twelve which is a high alpha-wave state, generally associated with closed eyes. Dr. Mikuriya's informal opinion about this rather surprising finding was that the verbal-rational left hemisphere had, in effect, gone into idle speed, allowing the intuitive, pattern-recognizing right hemisphere to take over.

TOUCHING THE WORLD

Start out by doing this exercise at a location from which you can see a tree or shrub. Stand and go through the balancing and centering procedure. Look in the general direction of the tree with soft eyes, then let the eyes focus on some particular leaf. With the index finger extended, let the left hand rise and point directly at that leaf. The arm and shoulders should be relaxed. Your assumption here is that some aspect of your finger actually *touches* the leaf. Perhaps you can visualize a beam of some sort of energy extending from the fingertip to the leaf, or perhaps you can experience the finger itself somehow transcending space to *touch* the leaf. The idea in any case is to assume mutual influence; by your act of intention, you are to some extent influencing the leaf and the leaf is similarly influencing you.

While your finger is touching the leaf, continue to be aware of your own center as the center of the universe. The leaf is also part of the universe that extends out from your center. During this exercise, due to the vector of interest expressed by your intentionality, that particular leaf is given a special significance.

After approximately five minutes, let your eyes go soft and swing your hand in a small arc a few degrees to either side of the leaf. Can you feel an increased sensitivity at the fingertip every time it touches the leaf?

Now drop your arm, shake out both hands, check your balance and center, and repeat the touching process with the same leaf. This time, if you can get a strong feeling of the leaf while moving your finger in the soft-eyed state, try the same thing with your eyes closed. If this is successful, try moving your finger even farther off the target, and check if you can find the leaf without opening your eyes.

The primary purpose of this exercise is to make it possible for you to sense a new connectedness with the world. At first, it is generally best to establish contact with living things or other organized entities

APPENDIX

(clocks, paintings, automobiles, etc.), but the exercise can be applied to any part of the world to which you are willing to give significance. Refinements can be added. One of my students, for example, has reported touching a hummingbird at a feeder, feeling the shape of its body, its pulsing throat, the vibration of its wings.

THE CRYSTALLINE STATE

In the crystalline state, there is no expectation nor any prejudgment. Concentration on the past and future gives way to primary focus on the present. Action taken while in the crystalline state is not *considered action* but rather *appropriate action.* When fully achieved, this state permits awareness of the perfect rhythm that always exists at the heart of your being. Going into the crystalline state does not require a set of instructions; it can come upon you spontaneously, as described in Chapter Eight. Likewise, no set of instructions can assure that you will achieve this state. The following procedure is offered as a guide. Your own intentionality is the key ingredient.

Sit comfortably, either in a meditation position or in a straight-backed chair. Balance and center. Using soft eyes, create an imaginary ball about the size of a volley ball between your hands. Hold the ball gently. Note that it is as clear and pure as a crystal. Sense its surface by moving your hands slightly together and apart in a rhythmic manner. Let the ball become as real as your intentionality. See if you can actually *feel* where its surface begins as you move your hands in and out.

Continuing to pulsate the ball, think of your most pressing personal problem, whatever is occupying your ego. Put this problem *in* the ball. Focus your eyes on the problem. Let it take on its own shape and texture and color. What does the problem look like now? Does it move? What part of the ball does it occupy?

Ask yourself whether you're willing to give up that problem for the next fifteen or twenty minutes. Are you willing to let go of it completely for at least this long? Assure yourself that you can get the problem back when the exercise is finished, if you so desire. (Your problems are important aspects of your ego; in this case, in fact, this particular problem may be said to represent your ego.)

Now, if you're really willing to give up your problem for the next fifteen or twenty minutes, press the ball with the problem in it down into the floor. Let it sink into the earth. The earth will serve as a bank; you can always recover the problem later.

Shake out your hands. Balance and center. Again create a ball between your hands. The ball is as pure and clear as a crystal. Focus your eyes on the center of the ball, allowing the background to go out of focus. Continue pulsating your hands. You might find it hard to stay focused on "nothing," but keep bringing your focus back to that central point. Continue this practice for at least five minutes.

During this period, consider that the crystalline state, now represented by the ball, contains no expectation and no prejudgment. It exists in the vibrancy of the present moment. In the crystalline state, what you experience is neither "good" nor "bad," neither "successful" nor "unsuccessful." It just *is*.

After five minutes or so of concentration have passed, ask yourself, "Am I willing to live during the next ten to fifteen minutes entirely in the present, without expectation or prejudgment?" If the answer that comes from the heart is no, just throw the ball away and, if you so desire, take back your problem from the earth. If your heart's answer is yes, however, take the ball and press it into your body, at the point of your center. This act represents your induction into the crystalline state. Feel the ball expand inside you and spread to fill your whole being.

In the crystalline state now, rise and walk around, experiencing familiar things in a new way—a chair, a table, a painting, a flower, a tree. Approach each experience without expectation or prejudgment. Go to a mirror and look at yourself from the intense clarity of the present moment. Perhaps you can arrange to meet a friend, a loved one, or a stranger while in this state. If so, there need be nothing out of the ordinary about your external appearance and actions; the crystalline state entails *appropriate action,* and it is appropriate to meet another person in a relaxed and natural manner. Your own experience during the meeting, however, will probably be transformed, and this transformation will most likely reveal itself in the essential quality of the relationship.

Whenever you wish, you can return to the state of expectation and prejudgment. Remember that your pressing personal problem is still there in the earth for you to retrieve and return to your consciousness, if you so desire. I have mentioned ten to fifteen minutes as an appropriate length of time for your first attempt at the crystalline state, but you can remain in it longer if you wish.

The crystalline state is an *alternative* mode of being. To offer it is not to denigrate other modes. The point is that to be able to move into the crystalline state at will, eventually without the rather involved induction procedure presented here, is to enjoy the possibility of a richer, more fascinating, and more humane life.

CRYSTALLINE EXERCISES

The crystalline state permits a temporary release
from the constraints of the ego, and thus allows the
possibility of experiencing the essential connectedness of all existence that has been described in this
book. The exercises here are based on that possibility.
Before attempting them, make sure that you are balanced and centered. Then enter the crystalline state
and maintain it during the course of the exercise.
Analysis and judgment should be withheld until each
exercise is finished. Bear in mind that you are the
authority for your own experience.

Scanning. Select some living thing or other organized entity, such as a tree. Extend your left hand and
arm in front of you, palm down. Aim it in the general
direction of the tree you've selected. With soft eyes,
sweep the arm through an arc that includes the tree.
See if you can feel the presence of the tree at a distance, as your arm passes. Practice this with soft eyes
until the feeling is distinct.

Close your eyes and disorient yourself by rotating
in place, first one way, then the other. Now, with eyes
tightly closed, scan for the tree by rotating the whole
body with the arm extended in front of the body. Rotate at the rate of about one full turn in five or six
seconds. When you sense the presence of the tree, go
past that place in the arc, then come back and zero in
by making smaller and smaller arcs. When your hand
and arm feel as if they are lined up with the tree, open
your eyes. Make no judgment about yourself, whether
your hand is on target or not. In terms of the crystalline state, the position of your hand is neither good
nor bad. It just *is.* Keep repeating the exercise. Try
different vantage points. Move far away from the tree.
Position yourself out of sight of the tree and see if you
can find it by scanning.

Now try the same thing with another person. If you can locate the person by scanning, have him or her move while your eyes are closed and try to locate the new position, or even follow the motion. Increase the distance. Try to locate a friend who is out of your line of sight.

There are obviously many exercises that can be done with this basic scanning technique. You might go to some natural setting, some forest or meadow, and simply scan for "something interesting." While doing this exercise, workshop participants have discovered such things as bee hives, hidden springs, and old grave sites. Another exercise used in workshop sessions involves scanning for magnetic north. Ideally, this should be done in an outdoor clearing on an overcast day. First, fix your intentionality on the idea of magnetic north, and concentrate on it for five minutes or so. Then extend your left arm and hand, palm down, and scan for magnetic north. *Do not try to figure it out.* Simply assume that magnetic north on this planet is part of the universe that is you. Try the exercise several times. Note which way your hand is pointing each time, without judgment. *After* the exercise, check your results with a compass.

Connectedness. After reading the description of remote viewing given in Chapter Six, you might work out a way to try something similar as an exercise, rather than as a scientific experiment, using the crystalline state. A simplified, informal version could involve only three people, call them Anne, Tom, and Carolyn, who would operate as follows:

At a predetermined time, say 10 A.M., Tom and Anne meet in a quiet room. At the same time, separately, Carolyn begins driving from some nearby point. She does not meet with Tom and Anne before beginning the drive that morning; this precludes their picking up any subtle hints from her as to where she is going. Carolyn drives anywhere within a half-hour's drive from where she started. By 10:30 at the latest, she has arrived at the site she has selected and is out of her car. At exactly 10:30, she goes into the crystalline state and, for the next twenty minutes, observes her surroundings.

From 10:00 to 10:30, Tom and Anne remain in the quiet room, relaxing, not talking. At exactly 10:30, Tom goes into the crystalline state and focuses his intentionality on Carolyn and what she is seeing. Anne does not go into the crystalline state, but begins asking Tom what he is seeing. (Bear in mind that neither Tom nor Anne knows where Carolyn has driven.) Tom simply gives any impressions he might have. He does not try to figure out where Carolyn is; he simply states his impressions—including shapes, colors, textures—without judging their quality or reasonableness. If he wishes, he makes sketches. Anne continues questioning him gently, adding nothing of her own. The conversation is tape recorded.

At the end of the twenty-minute period, Carolyn drives back and meets Tom and Anne. She listens to the tape and gives feedback. Then the three of them drive to the site that Carolyn had selected. On other occasions, the three take turns playing each of the roles.

It is important to stress that the exercise presented here is not shaped as a scientific experiment. Whatever the outcome, it proves nothing in scientific terms; it makes no claims. What it does is offer practice in enhancing your sense of connectedness with others.

Tuning-In. This exercise requires a partner. You and your partner sit facing each other in the meditation position or in straight-back chairs, so close that your knees touch. Balance and center. Both you and your partner should go into the crystalline state. Place your hands on your knees, palms up, and have your partner place his or her hands, palms down, into yours. Both of you should then close your eyes.

Assume that your partner's bodily states are known to you at the deepest level, and that, in fact, you will feel in *your* body every bodily state your partner feels. Tune-in to your partner's state for a few minutes. Then, if you feel a tightness in your own neck, say aloud to your partner, "There's a tightness in your neck. See if you can relax it." Keep tuning-in. If you should feel your neck relaxing, say aloud, "Good, your neck is relaxing now." Keep tuning-in by mentally scanning your own body for anything out of the ordinary. Report anything you feel to your partner as if the condition exists in his or her body.

After all possible corrections have been made, start working on positive change, on creating a sense of glowing aliveness and awareness in your partner's body. Again, use feelings in your own body as a guide; for example, "Your head feels full of life and energy, but the aliveness seems to stop at your neck, leaving your body rather numb. See if you can let the energy flow downward from your head into your body." Keep tuning-in. When you feel a change in your body, report it as a change in your partner's body. "Good. Now the energy has flowed down through your neck as far as your heart. See if you can let it flow downward and fill your whole body." Continue giving instructions in this manner until your own body feels relaxed, alive, and glowing. Then come out of the crystalline state along with your partner, and compare notes. Which of your instructions were on target? Which ones were meaningful and useful? Now change roles and repeat the exercise.

Feeling for Others. This exercise applies the techniques learned thus far to a broader social context. In a recent newspaper, magazine, or book, find a picture of someone from a social milieu quite different from your own, say an African or Indian villager who is threatened with starvation. Learn everything you can about the villager's situation. If possible, read not only the article in which the picture is included, but also other articles and books on the subject from various points of view. Note the complexity of the situation on the cognitive level. Consider in what ways this one villager's plight is connected with political, social, technological, and economic matters that reach all around the globe.

When you are as well-informed as possible, pick a period during which you can be alone and uninterrupted, and go into the crystalline state. This means temporarily adopting an alternative way of being and knowing. Your assumption here is that the villager is part of the universe that is you, and that you can choose to experience that part of you directly; that is, you can *be* the villager. Look deeply into the picture, then close your eyes. Be willing to relinquish the past and future and all such considerations as guilt or pity. Accept the purity and clarity of the present moment, the point of the silent pulse where existence unfolds and all things are connected. Tune in to the villager's being. Experience existence *as* the villager. Stay with the experience, if you can, for at least a half-hour.

Return to the normal state of consciousness. Now take what you have learned cognitively and synthesize it with what you have experienced in the crystalline state *as* the villager. Note that the two states are not contradictory but complementary. How does the new synthesis affect your knowledge and insight on the situation? How does it affect your stance towards the larger problem, your willingness to act in the world?

A SYNCHRONIZATION PROCESS

This process is presented as an example of the kind of large-group activity that is carried on at energy-awareness workshops. These workshops are devoted not only to enhancing human communication and connectedness but also to teaching alternative ways of dealing with conflict and stress. This particular process is suitable for groups of 20 to 200 in a space such as a hotel ballroom, gymnasium, or large meeting room. There are numerous other processes that make up the workshops, some of which have been used with groups of more than a thousand people.*

The leader begins this process by asking the participants to clear the floor of all extraneous objects, then spread out and stand facing him so that each of them can turn all the way around, arms extended, without touching anyone else. Participants are led through the balancing, centering, and soft eyes procedures. Then they are encouraged to move swiftly and at random all around the room. Soft eyes make it possible to do this, even at a rather fast pace, without colliding with anyone else.

When the group is thoroughly mixed, the leader asks everyone to stop, close eyes, and recheck balance and center. "Somewhere in the room," he says, "there is a partner who is appropriate for you to work with during this process. Standing in place, extend your hands and, leaving your eyes closed, rotate until you sense your partner's location. Then move slowly, eyes closed, sensing your way through your hands, until you find a partner. Take hands with your partner, leaving your eyes closed, and get to know her or him by making subtle body movements. These movements

* For information about energy-awareness workshops, write George Leonard, P. O. Box 509, Mill Valley, Calif. 94941.

will be sensed through your hands. What can you learn of your partner's rhythmic essence just through these movements?"

Generally a few people in these large groups are unable to find partners with their eyes closed. The leader or his assistants help these people get together.

"In a moment," the leader continues, "I'm going to clap my hands. When I do, open your eyes and immediately close them again. Take a quick snapshot of your partner, in which you get only the essence of the configuration of the face and form. Let this image, this essence, dissolve within you. Let it become part of you."

Two or three minutes later, the leader continues: "Now open your eyes and watch this demonstration with soft eyes." He walks around the room stride for stride with an assistant or one of the participants— shoulder to shoulder, arms linked. "Walk this way with your partner, as if the two of you are a single organism. Soft eyes. Keep walking until your movements are completely synchronized."

For five minutes or more, the couples walk around the room. The leader reminds them to consider themselves a single physical organism with their partners. Then he passes out sheets of paper, one to each couple, and tells them to spread out to the edges of the room and sit facing outward, each couple close together. He asks that they read aloud from the sheet of paper, synchronizing their words with their partners' words (but not with the words of the other couples). The written material can be an excerpt from any reasonably rhythmic prose work. The leader suggests that the participants read not for meaning but just for rhythm, and that they assume, again, that they are one with their partners. When they come to the end of the page, they are to start over from the beginning.

Soon there is a pleasant hum of sound throughout the room as the couples read aloud, each couple at their own speed and rhythm. During this part of the process, the breathing of most of the couples becomes synchronized, and perhaps the heartbeats and brain-waves as well.

After five to ten minutes, the leader tells the couples to leave the sheets of paper where they are sitting and begin the synchronized walking again. The cycle is repeated twice more—walking-reading, walking-reading—after which the couples are asked to sit facing each other, knees touching. Each couple then makes a single large crystalline ball between them; four hands pulsate the ball. The leader reminds them of the qualities of the crystalline state. He asks the participants to focus their eyes on the center of the ball, then raise it to eye level.

"Please keep your eyes focused on the center of the ball," the leader says. "This will mean that your partner's face will look out of focus to you. Just keep trying to hold your focus on the center of the ball. . . . Now, there's one characteristic of the crystalline ball we haven't yet discussed. It resonates only to a single frequency. In this case, we're going to assume it's tuned to a frequency that represents a resonance common to the two of you, a key frequency in this new unified organism. When I clap my hands, I ask that you look directly at your partner's eyes *through* the crystalline ball. Let your eyes meet through the ball, on a single, clear, resonant frequency."

The leader claps his hands, and the partners look directly into each other's eyes for some five minutes. Then the leader asks that they crush the crystalline ball out of existence between their hands and rise to a standing position, now in the crystalline state.

"In a moment, I'm going to ask that you say goodbye temporarily to your partner. Though you'll be physically separated, you'll still be joined in some very significant manner. The assumption is that you'll always know exactly where your partner is. I'm going to ask that you start the random, soft-eyed walk that you did at the beginning of this process. Then, when I clap my hands, stop, extend your hands, rotate in place, and scan for your partner. Don't *look* for your partner. Keep your soft eyes. When you sense your partner's presence through your hands, zero-in and stop, facing your partner wherever he or she may be in the room."

The random walk begins. When the leader feels that the group is thoroughly mixed, he claps his hands, the participants stop, rotate in place, and zero in on their partners. This is repeated with soft eyes, as a practice for what is to come. The participants are then asked to go through the same procedure with eyes barely open, so that they can see only the feet of nearby people. This, too, is repeated.

Finally, they are asked to walk around with eyes just barely open, then to close the eyes entirely at the sound of the leader's hand-clap. Scanning for partners takes place, in this case, with eyes completely closed. When the participants are zeroed-in, the leader asks that they open their eyes and, if necessary, correct their alignment. The closed-eye practice is repeated several times. The group then gathers informally for an open discussion of the process.

A BRIEF EDITORIAL

No matter now extraordinary the results obtained in such processes, the ultimate purpose of this work is not to demonstrate spectacular new powers; it is to increase harmony and justice in the world. Human capabilities are obviously much greater than we generally assume. As we begin to realize some of these capabilities, we move into fascinating realms, full of their own possibilities and dangers. What will count in the long run is not just what we learn to *do* but what we are willing to *be*. The most promising adventure is worth joining only if it eventually contributes to the common good.

INDEX